T0191402

WEARABLE
MEDICAL
TECHNOLOGIES

KEVIN CHEN

WEARABLE MEDICAL TECHNOLOGIES

Books Beyond Boundaries

ROYAL COLLINS

Wearable Medical Technologies

KEVIN CHEN

First published in 2021 by Royal Collins Publishing Group Inc.
Groupe Publication Royal Collins Inc.
BKM Royalcollins Publishers Private Limited

Headquarters: 550-555 boul. René-Lévesque O Montréal (Québec) H2Z1B1 Canada
India office: 805 Hemkunt House, 8th Floor, Rajendra Place, New Delhi 110 008

ISBN: 978-1-4878-0489-3

To find out more about our publications, please visit www.royalcollins.com

Foreword

"Recently, the phrase 'draft of wind' has been trending on the Internet. I believe that riding on the draft of wind of 'Internet Plus' will allow the Chinese economy to soar." This was what Premier Li Keqiang said at the press conference of the Third Session of the 12th National People's Congress in 2015. After it was quoted by the Premier, Internet Plus became a strong draft, and a huge wave of industries wanted to get in on it at once.

Since the emergence of the Internet, many industries have begun to seize this opportunity and transform, by getting involved with Internet Plus. After Premier Li Keqiang mentioned it, Internet Plus was instantly elevated from a "folk remedy" to a national strategy. One can imagine the number of policy dividends this brought about as a result.

Today, in the 21st century, the Internet has become a form of infrastructure, while "mobile Internet" is beginning to subvert all mankind's way of life. People's clothing, food, living, and travel behaviors have undergone profound changes due to mobile Internet. Socializing methods, in particular, are constantly being upgraded with the introduction of various apps.

In a country like China, which has a huge population base and uneven resource allocation, access to a doctor has gradually become a pressing issue. Although the central government's medical reform is becoming more thorough and achieving some breakthroughs, the problem whereby "medical treatment is expensive and difficult" remains acute.

For the general public and for the longest time, the thought of seeing a doctor has been a cause for headaches. But for enterprises, investors, and entrepreneurs, medical treatment has always been a blue ocean full of business opportunities. As medical treatment combines with mobile Internet, the era of mobile health care is fast approaching.

According to an analysis report published by Sootoo Research, 2015 will be the year during which mobile health care will experience exponential growth. China's mobile health care market is expected to amount to 4.5 billion *Yuan* that year, 8 billion *Yuan* in 2016, and 13 billion *Yuan* in 2017. The global mobile healthcare market is expected to amount to USD 23 billion in the same year.

It is evident from this data that the outlook for mobile health care is rather promising. It is a shining "gold mine." Whoever seizes the opportunity first will gain an absolute advantage. Looking at the big picture, you may find opportunities everywhere in mobile health care, but that does not mean it is easy to find a point of entry. Hence, through this book, the author will show you that the best point of entry to mobile health care is the "wearable device."

Why is this so? The author will answer this question in this book.

Contents

CHAPTER 3

Current Status of Wearable Health Care   /   45

CONTENTS

CONTENTS

CHAPTER 1

Wearable Devices

In April 2012, Google invented a type of smart glasses, which triggered the concept of wearable devices. This pair of glasses has an augmented reality function. It can also take photos, make video calls, and identify directions via voice control. Furthermore, it has novel functions such as surfing the Internet and processing text messages and emails.

1.1   Concept Trigger

Wearable devices, as the name suggests, refer to various types of smart devices that can be worn on the body. These devices have all the functions that are present in today's smartphones, tablets, and PCs, but the biggest difference lies in the presence of various high-precision and sensitive sensors embedded in them. These input terminals can achieve unprecedented deep integration with the human body. For example, input methods are no longer the traditional keyboards or voice activation, but upgraded forms that are closer to the human body, such as heartbeat, brain waves, and retina, etc.

When wearable devices have developed to become sufficiently mature, they will become part of human life or even the human body. For example, they can be in the form of anything that is closely related to people's daily life, such as glasses, wristbands, watches, clothing, and footwear. The trend of the future development of wearable devices is a replacement of mobile phones, which everyone owns today, to become a new data traffic portal with no time limit and no boundaries.

At present, the most common wearable devices in the market include smart glasses, represented by Google Glass, smartwatches, with Apple, Samsung, Sony, and Pebble in the lead, and smart wristbands, whose market is mainly occupied by Fitbit, Jawbone, and Nike. Besides, there are virtual reality headsets mainly developed for games, such as Oculus Rift and Sony's Project Morpheus.

Every part of the human body can be a potential field for the development of wearable devices. It is not limited to prominent places such as the head, wrists, and feet. There are many micro-sensing devices implanted inside the human body that are making breakthroughs in new fields and bringing about a huge revolution in the entire healthcare field.

1.2   Forms of Devices

The main wearable devices in the market today come in various forms, including smart glasses, smartwatches, smart wristbands, smart running shoes, smart rings, smart armbands, smart belts, smart headsets, and smart buttons, etc. (Figure 1-1) Among them, the development of smart wristbands is the most mature, followed by smartwatches and smart glasses. In addition to these three, user demand for devices in subdivided categories has also been on the rise. These include jewelry and embedded sensors for T-shirts and shoes. In 2015, Gartner (an IT research and consultancy company) predicted that the shipment of smart clothing will jump from 100,000 in 2014 to 10 million, which is equivalent to 1/3 of global smartwatch sales.

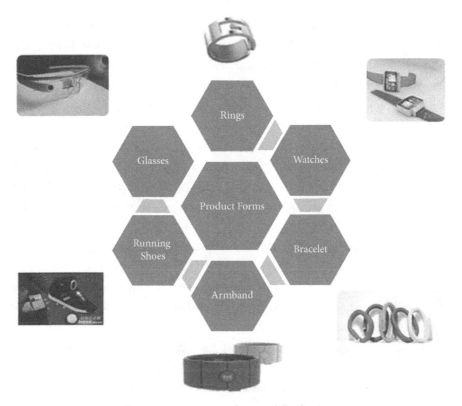

Figure 1-1 Forms of wearable devices

According to predictions made by the Internet Data Center (IDC), the total shipment of wearable devices is expected to amount to 112 million by 2018, which is more than four times higher than in 2014.

1.2.1 Glasses: Google Glass

Google Glass (Google Project Glass) is a type of augmented reality glasses released by Google in April 2012. It has the same functions as a smartphone. The biggest difference lies in Google Glass' ability to take photos, make video calls, and identify directions, as well as surf the Internet and process text messages and e-mails via voice control.

3

A camera is suspended at the front of the Google Glass, and a wide strip-shaped computer processor device is located on the right side of the glasses. The equipped camera has 5 million pixels and can shoot videos in 720p. The lens is equipped with a head-mounted micro-display that can project data onto a small screen above the user's right eye. The display effect is equivalent to a 25-inch high-definition screen 2.4 meters away.

At present, applications of Google Glass that are relatively more distinctive include: analysis of emotions through expressions, broadcast of NBA matches through the Google Glass, assisted teaching, and medical research. Among them, the advantages gained from the applications in medical research and teaching are the most prominent.

According to FierceHealthIT, a website for medical and healthcare information in the United States, several U.S. healthcare industry experts said that a growing number of developers are exploring new ways to use the Google Glass technology within the healthcare industry, in a bid to bring about revolutionary changes in the field.

Bionic-wise, another medical and healthcare information website in the U.S., also published an article saying that Google Glass will bring about innovations in the technology fields across the healthcare sector in at least eight ways, such as:

1. *Google Glass apps developed especially for patients with Parkinson's disease.* Researchers at the University of Newcastle in the United Kingdom are developing a new idea to help Parkinson's disease patients live more independently through Google Glass' automatic reminder function. For example, when Parkinson's disease patients are outdoors, the app can maintain real-time contact with the patients' relatives and friends, and also remind patients to take medication, allowing them to better take care of themselves.

2. *Access patient images and information online.* Taking OneDx, a software platform, for example, this platform allows physicians to access information such as medical reports, the location of inpatients, and laboratory information using Google Glass. Doctors can download materials such as

research results, reports, and laboratory sheets during their breaks amidst a busy workday.

3. *Smart recording and identification of patients' information.* At present, six clinics in the U.S. are using the Google Glass software developed by Augmedix. When doctors are communicating with their patients, the software can automatically electronically enter the patients' information. In addition, the video capability of Google Glass allows the software to identify the painful areas which patients are referring to, based on nonverbal communication with the patients.

4. *Used for medical education and training.* In the operating room, doctors can use Google Glass to present footage of operations to students in a first-person point-of-view, so the latter can see the actual situation even if they are not physically standing beside the operating surgeon. When Dr. Christopher Kaeding of Ohio State University Wexner Medical Center was performing knee surgery on a patient, with the patient's consent, he live-streamed the operation through Google Glass.

Although it is difficult for Google Glass to be popularized among the masses, so much so that it once withdrew from the field of wearable devices, all present signs show that it can better fulfill its potential in the professional field. The most promising field will be the medical industry.

1.2.2　Watches: Apple Watch

The Apple Watch is a type of smartwatch released by Apple in September 2014. It has three models, namely the ordinary model (Apple Watch), sports model (Apple Watch Sport), and customized model (Apple Watch Edition). It supports phone calls and voice messages replies, can connect to cars, obtain weather and flight information, and has dozens of other functions such as map navigation, music playback, heartbeat measurement, and step count. It is a comprehensive health and exercise tracking device.

On 2 June 2014, Apple released a new mobile application platform at the Worldwide Developers Conference. Named Healthkit, this platform can collect and analyze users' health data. Apple's executives told developers that it can organize data, like blood pressure and weight, collected by various other health apps on the iPhone, iPad, and Apple Watch.

In 2015, after the release of the Healthkit, Apple created a software infrastructure specifically for medical researchers. Its main objective was to resolve the present difficulties faced during medical investigations, such as a lack of samples and participants, and insufficient data support.

The greatest value of this platform lies in its ability to assist researchers and medical practitioners in collecting and organizing patients' medical data, and in helping people diagnose various diseases, using the iPhone. Its applications now cover breast cancer, diabetes, Parkinson's disease, cardiovascular disease, and asthma. Users can monitor and track data about their physical conditions with the Apple Watch.

Cook also officially stated at the press conference that using the Apple Watch can make a person healthier. Apple Watch's Activity app can display the user's amount of daily exercise, including calories burnt by walking and the amount of time spent exercising. This information can be promptly fed back to the user. Besides, users can also set reasonable goals, and the Apple Watch will remind users daily according to the settings to assist them in achieving their goals.

1.2.3   Wristbands: Jawbone UP

Jawbone UP (Figure 1-2) is a smart wristband released by the famous Bluetooth headset and speaker manufacturer, Jawbone, in 2011. Jawbone UP is a wristband device that can track information such as users' daily activities, sleeping conditions, and eating habits. It has a smart alarm, idle alert, special alert, power nap mode, and other application modes. Jawbone UP is designed to integrate seamlessly into people's lives.

Its health tracking function makes use of various bioimpedance sensors to gather data about the heart rate, respiratory rate, skin resistance to electric current, skin temperature, and the ambient temperature of the user. In addition, the device can more accurately display a user's step count, distance covered,

Figure 1-2 Jawbone UP

speed, and calories burnt during exercise, all according to information input, such as the user's weight, height, age, and gender.

Not long ago, Jawbone upgraded the smart wristband's physical recording application. In the original model, users were required to add dietary information over and over again, by tapping on the screen for about 15–20 times. The upgraded application is more user-friendly and makes it easier for users to calculate calories. Moreover, the upgraded application can also be connected to the database for restaurant menus, which contains calorie information for each dish served by the restaurants, as well as information on nutritional value submitted by UP users.

Dietary management is also a unique application of Jawbone UP. Other sports wristbands are only capable of the same old functions such as step counting, calorie burn calculation, and sleep quality analysis, which have caused the field to become almost completely homogeneous. This application of Jawbone UP is thus of great significance, as it has broken through the current tepid growth of sports wristbands in a certain sense, and it has become the first to enter the phase of real big data application and platform construction.

1.2.4   Jewelry: Smart Ring Ringly

The startup Ringly designed a smart ring specifically for women who want to look good (Figure 1-3). The Ringly smart ring can be connected to mobile phones with iOS and Android systems via Bluetooth. As many women may miss important calls or text messages while their mobile phones are in their bags, this smart ring comes in handy by alerting its users.

Users can set different vibration modes and five different colored alert lights according to the identities of the caller, such as a spouse, boss, and close friend. This function prevents the Ringly ring from constantly vibrating and flashing.

Unlike other smart devices, this smart ring achieved a breakthrough in design, as it could, to a certain extent, integrate technology and fashion.

Figure 1-3 Ringly

According to a survey conducted by The Journal of the American Medical Association, the middle class in the U.S., which earns an annual income of more than USD 100,000, is generally not interested in wearable devices that are too complicated and geeky. Perhaps this smart ring, which does not look too geeky or complicated, will be able to draw the attention of these middle-class groups.

1.2.5   Virtual Reality Headset: Oculus Rift

On 1 August 2012, Oculus Rift, a Virtual Reality (VR) headset that appeared to be part of science fiction but was actually scientifically-backed, was put up on the crowdfunding platform Kickstarter by Oculus, awaiting public investors to make their "honorable visits." Oculus proclaimed that the Oculus Rift would "change the way you think about gaming forever" in its crowdfunding pitch. Looking at the performance of this device in the gaming industry today, it is clear that the Oculus Rift has lived up to expectations.

Oculus Rift is a VR head-mounted display designed specifically for video games. It instantly captured the hearts of the public with its unique features. In just one month, it won the support of over 9,000 consumers and received USD 2.43 million of crowdfunding funds, which served as the initial sum of money accumulated for its subsequent development and production.

Oculus Rift is equipped with two eyepieces, each eyepiece with 640 × 800 resolution. Both eyepieces combined provide a binocular vision of 1280 × 800 resolution. The biggest feature of this device is its perspective which is controlled by a gyroscope, as this greatly enhances the player's immersion in the game. Oculus Rift VR glasses can be connected to computers or game consoles via DVI, HDMI and micro USB ports.

On 26 March 2014, Facebook announced its acquisition of Oculus VR, the immersive VR technology company, for approximately USD 2 billion, thereby officially entering the field of wearable devices. Subsequently, in July 2014, Facebook announced another acquisition, of RakNet, the game development engine. Facebook then turned it into a free and open-source resource for game developers around the world. This brought Oculus Rift to a new phase of development – platform construction.

VR headsets that are similar to Oculus Rift, such as Sony's Project Morpheus and Samsung's GearVR, were mainly developed for the gaming industry in their initial stages. But following further development, VR technology has gradually shown unique advantages in other areas, such as the medical industry.

The technology behind VR works by constructing a virtual natural environment using computers and professional software. Such an environment can assist doctors in carrying out a series of realistic training such as disease diagnosis and rehabilitation training. For example, during training, doctors can simulate a patient's condition with a pre-programmed case. That simulated patient will respond virtually according to the doctor's actions, whether they are right or wrong.

In fact, a French surgeon had worn a VR device during hip surgery, so that other junior doctors could watch the video for experiential learning. Oculus Rift and GoPro were used together as teaching tools for surgery. During the operation, two synchronized GoPro cameras were installed on the surgeon Gregory's forehead. They were responsible for recording Gregory perform the hip replacement surgery from a bird's eye view. In the end, a 3D surgical video from Gregory's perspective was created.

This video can be played for medical students and junior doctors using the Oculus Rift. This way, medical students and junior doctors can watch the surgery from the perspective of the surgeon. This is unprecedented.

In addition to more effectively helping doctors complete their training in the medical industry, there is a new function for VR technology, which is to put a person in another's shoes. The environment simulated through VR technology allows users to experience the misfortunes of others to the maximum extent, and to perceive their joy, anxiety, and pain.

Oculus Rift can solve the problem of Telepresence to the maximum extent, which will virtually place one in a location beyond the reach of his or her physical body, and allow one to have a real experience beyond space, time, and medium. The moment you put on the Oculus Rift, you will slowly enter the other party's world, and even become the other party. Gradually, you will question your true identity and break down your inherent gender cognitions.

The future development of VR technology is not limited to the gaming industry. It will penetrate various fields and play an increasingly important role, especially in the medical and tourism industries.

1.2.6   Clothing: Athos

We often define wearable devices as various types of smart devices that are worn on the body. In fact, more accurately speaking, they are the various sensors that are worn on the body. Without sensors, smart wearable devices become worthless, but with sensors, they transform into powerful devices that can wow people at any moment.

For an ordinary piece of clothing, we may pay more attention to the style and price, but if a sensor is added to the clothing, that will be a completely different matter. Athos smart fitness apparel is one such example. It has a large number of built-in sensors in different places, all placed according to the human body structure. These sensors can monitor the user's heart rate, respiratory rate, and even muscle activity using EMG sensors.

Athos smart fitness apparel contains an Athos Core chip and multiple sensors that can send data to smartphones via Bluetooth. This small device weighs less than 20 grams and can be used continuously for more than 10 hours. In addition to sending data, Athos can also monitor the user's exercise condition through a built-in six-axis accelerometer.

Without sensors, such apparel would be no different from ordinary clothing. It is because of these sensors that users can have more fun during exercise, and can see their various quantified health statistics.

1.2.7   Footwear: Lechal

Lechal is a type of smart footwear with a GPS function. It connects to mobile phones via Bluetooth. As long as Lechal is connected to Google Maps, the user can input his or her destination in their phone, and Lechal will alert the user of their path using vibrations. Besides, the Lechal smart footwear is equipped with a variety of health sensors, which can record the user's walking distance and calories burned at any time.

Krispian Lawrence, the co-founder and Chief Executive Officer (CEO) of Ducere said, "In Hindi, Lechal means to 'take me along.' The original intention of designing this footwear was to help the blind so that they can easily avoid walking into obstacles. Particularly in environments that they are not familiar with, this shoe will be able to play an important role."

1.2.8   More Diverse Forms

At present, most wearable devices mainly take forms of watches or wristbands. The potential for wearable devices has not been fully developed. As mentioned in the previous definition of wearable devices, wearable devices are essentially wearable sensors. With the emergence of new sensors in the future, diversification of their forms will be the major trend in the development of wearable devices.

At present, most wearable devices only exist as accessories for smartphones, and basic information transfer between the two devices can only be made using wireless communication technologies such as Bluetooth or WiFi. For example, in the case of some smart wristbands, information can only be viewed through the smartphone app after it is transferred from the user. This limits the scope of application of the wearable devices to a large extent.

Smartwatches, smart wristbands, and smart apparel are merely products at their initial stages of development within the wearable devices field. Future wearable devices will be fully integrated into the user's body without notice, and will naturally become a part of it. For example, Northwestern University and the University of Illinois have jointly embarked on a new wearable device project. The device itself is soft and can be attached to the skin like a tattoo. It can monitor body temperature, blood flow, and skin moisture, and transfer data directly to the user (Figure 1-4).

Moreover, most wearable devices have relatively limited applications, an accumulation of functions, and problems of homogenization. Most are only capable of informing users of the calories burnt each day, their heart rate during exercise, blood pressure, and blood glucose levels, among other factors. Perhaps they will be able to monitor the surrounding air quality and inform

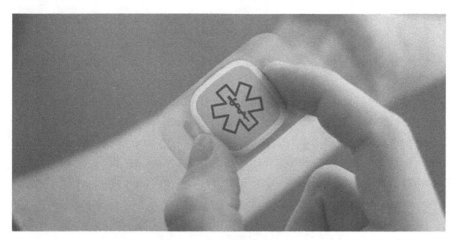

Figure 1-4 Wearable tattoo

the user whether to wear a mask out or not in the future, too. But from a macro perspective, wearable devices that are created solely for health monitoring are only concentrated in the consumer field. In reality, wearable devices can do more for humans and bring us more surprises.

Particularly in the medical field, wearable devices can fulfill their maximum potential for the benefit of humanity. For example, Khushi necklace, a crowdfunded device, can receive vaccination data through an NFC chip to improve the vaccination rate of children. A type of smart glasses developed by the OrCam company can transmit surrounding information such as letters, human faces, objects, products, locations, bus numbers, and traffic light signals to visually impaired users to help this group of people live more independently. CxExplorer, a type of smart sunglasses developed by the start-up EnChroma, can help people with color blindness better distinguish colors.

Furthermore, according to the McKinsey Global Institute, wearable devices have great potential in the area of medical preventive care. By continuously monitoring and synchronizing data with doctors, wearable devices can reduce patients' cost of treatment by 20%. Also, wearable devices can play an effective role in Post-Traumatic Stress Disorder (PTSD), mood swings monitoring, and even early cancer prevention.

1.3   The Smart Key Which Connects People and Things

No matter how popular smart homes, smart cities, the Internet of Things, or big data industries may be in the future, they cannot replace wearable devices. These industries have one thing in common, which is that they are only targeted at the connection between objects, and only work within the intelligent connection and informatization between objects. However, for these industries to further develop and to realize the ultimate concept of human-serving intelligent technology, they must rely on wearable devices to bridge the connection between objects and people.

Many people have a misunderstanding about wearable devices, as they often view wearable devices as pets of technology enthusiasts. Such an understanding deviates from the true meaning of wearable devices. From my understanding, the biggest value of wearable devices, which is also the core value that distinguishes them from smart homes, smart cities, or the Internet, lies in the fact that they are the only devices that can create and realize the connection between people and smart hardware in the era of mobile Internet.

There is some truth in the saying, "Smart wearable devices will replace mobile phones to become the center of the world in future." Although current mobile applications, as well as some applications of smart wearable devices, can only be used with mobile phones, we cannot assume that wearable devices can only go as far as becoming a mobile phone accessory.

There is hardly any difference between hardware like mobile phones and smart homes. It is just that mobile phones serve as a communication tool, so we have given it more functions, and at the same time, we have allowed ourselves to be inseparable from them in a virtual and contrived way. But the core difference between a wearable device and a mobile phone lies in the digital connection between people and objects. This is something that mobile phones cannot achieve, and the same goes for the fields of smart homes, smart cities, and the Internet.

Wearable devices can be truly implanted into and bound to the human body, where it can recognize the body's characteristics, and digitize and quantify everything in the human body. Therefore, be it smart cars, smart homes, smart cities, or the Internet of Things, wearable devices are a smart key that is required to connect with people effectively.

The value of wearable devices does not merely lie in its position as a new value entry point for mobile Internet. In the next wave of businesses, smart, mobile Internet, big data and other industries can only rely on the smart key, wearable devices, to unlock the connection to the human body if they wish to connect with people and gain commercial value for solving problems for people. Wearable devices are irreplaceable in this aspect.

Here, wearable device should be redefined as such, "wearable device is the key to connect people and smart devices". This definition will affect the industry's understanding of wearable devices, as well as the consideration of the true value of wearable devices.

1.4   The Best Entry Point for Mobile Medical Hardware

Research by the market research agency, Transparency Market Research, shows that the medical industry is the most promising industry for the application of wearable devices, followed by the fitness and entertainment industries. Ahadome predicts that wearable technology will account for at least 50% of wearable devices in the healthcare sector.

Wearable devices will bring a revolutionary change to the medical device industry, from miniaturization to portability, and eventually to wearability. Not only can wearable devices monitor health indicators such as blood glucose, blood pressure, heart rate, blood oxygen content, body temperature, and respiratory rate in real-time, they can also be used for the treatment of various diseases. According to data from iiMedia Research, the size of China's wearable medical device market is expected to reach 1.2 billion *Yuan* in 2015 and 4.77 billion *Yuan* in 2017, with a compound annual growth rate of 60%.

The best entry point for mobile medical hardware is none other than wearable devices. Its advantages include:

1. *It has inherent advantages in user training.* Before the emergence of mobile Internet and the appearance of wearable devices in the true sense, there were already many devices such as blood pressure meters and blood glucose meters in the market. According to statistics, these devices can achieve millions of units in domestic sales every year.

   User demand for such equipment will continue to increase. Besides, the function of wearable devices is no longer limited to measuring blood pressure or blood glucose levels. They can further manage users' health, and this will further stimulate the market potential of simple medical measurement equipment, as users will all move towards wearable devices with more complete and personalized functions.

2. *It truly liberates the users' hands.* The biggest advantage that wearable devices have in replacing smartphones is that it can truly liberate users' hands and become a new input and output terminal for the mobile network. The user does not need to input information through any of the traditional ways like typing, moving, or producing sounds. They only have to put on the wearable device, like wearing the smartwatch on the wrist, and the input will be done. Users' heart rate, amount of calories burnt, blood glucose levels, and brain waves have become the input signals for wearable devices.

3. *It can constantly monitor users' bodies.* As long as the user's body is functioning, the wearable device will be able to continuously record and analyze data through built-in sensors. Perhaps the current wearable devices, such as glasses, watches, and wristbands, may often be taken off by the user and thrown aside, but future wearable devices will be developed to naturally integrate with the human body. So much so, that their presence will not be felt. For example, they may be embedded in all clothes, shoes, and even the human body.

4. *A "chemical reaction" is taking place.* The concept of mobile healthcare has spread far and wide, but it has not yet been truly popularized. The present mobile health care is largely realized by a single combination of apps and smartphones. Strictly speaking, this is not truly mobile health care, because the most important data value is far from utilized and used.

The ongoing "chemical reaction" refers to the process whereby sensors constantly achieve breakthroughs, conduct integrated innovation and become deeply integrated with advanced technologies such as the Internet technology, mobile Internet, cloud storage, and big data analysis. Wearable devices are the best carrier for the output of this chemical reaction.

When all these technologies have become deeply integrated, the problem of data value mining will be naturally solved. In the future, smart wristbands will be able to keep track of various physical signs of the human body. This data will be first recorded by the device, then analyzed and fed back through supporting infrastructure in the ecosystem such as the Internet, cloud, and mobile health platforms, to provide users with targeted health advice to prevent diseases, or recommend corresponding medical resources for users with medical needs.

The entry point for the Internet in the era of wearable devices will not only be limited to a narrow scope. Any place where people go and anything they touch can be an entry point. Moreover, the information content brought by these entry points, and its accuracy will far exceed the traditional entry points of the Internet.

As mentioned above, the narrow definition of wearable devices is wearable sensors. Users use wearable devices not because of the functions that are available on smartphones, but because of the data collected by the sensors scattered over various parts of the human body. With an in-depth analysis of this data, users can accurately understand their physical conditions without a doctor and can make timely adjustments to their bodies.

The mobile health care in the true sense is not about allowing users to communicate with their doctors and be diagnosed in any location, but about users being able to diagnose themselves on a deeper level, anytime, anywhere. In the future, all doctors and patients may even be required to diagnose patients' conditions based on the same data analysis platform, which would be a big data platform that consists of data recorded by wearable devices and analyzed by cloud computing.

CHAPTER 2

The Healthcare Situation in China

2.1 Existing Disadvantages

2.1.1 Concentrated Medical Resources, Patients seeking Treatment from Large Hospitals for Minor Illnesses

With the development of the national economy, there has been a rise in people's standard of living, as well as the demand for medical resources. As a result, the conflict between the demand for health services and the availability of medical resources have become increasingly prominent.

According to the data released by the Ministry of Health, 80% of China's medical resources are concentrated in large cities, and 30% of these resources are distributed across large hospitals. There is an uneven distribution of medical and healthcare resources across different regions. Even in the same region, there is a huge difference in distribution among hospitals of different tiers. Moreover, villages and small communities in cities lack qualified health personnel and general practitioners, and even some small and medium-sized hospitals in cities lack highly-skilled doctors. This has caused people to think that they should flock to large cities and large hospitals to seek medical treatment, whether for major or minor illnesses.

This in turn leads to a phenomenon whereby there are endless registrations made to see some specialists in these large hospitals every day, and yet the patients may have merely caught the common flu. This is also a hindrance to the improvement of doctors' medical skills, as a considerable amount of time is spent on patients whose illnesses are not related to their specializations. Large hospitals are supposed to treat patients who are critically ill or suffer from conditions that are difficult to treat, but instead, they have been attending to a large number of patients with common and frequently-occurring illnesses. This not only makes it difficult and expensive for the masses to seek medical treatment but also wastes many precious resources.

Take Beijing as an example, it is known as the "National Medical Center." A large number of people from other places travel to Beijing for medical treatment, and this has overwhelmed the Tier 3 hospitals in Beijing. According to statistics, an average of 700,000 people travels to Beijing from other places for medical treatment daily. The neighboring Hebei Province contributes 7 million visits annually, with its people insisting on traveling to Beijing to seek medical treatment even for colds and fevers.

2.1.2   Seeking Medical Treatment has become more expensive

The data in the *Annual report on reform of medical and health system in China* published in November 2014 shows that, although the proportion of personal health expenditure to the total health expenditure in China fell from 40.4% in 2008 to 34.4% at the end of 2012, personal expenditure has risen by 64.31%. The most intuitive point that can be drawn from this set of data is that the issue brought about by the previous medical reform was not about its rising medical fees, but an increase in the proportion of medical expenses borne by people. In other words, it has become more expensive for the general public to seek medical treatment.

The train of thought behind the medical reform was not to effectively increase the total supply of medical services but to control the price of the medication.

It was to reduce the price of medication, strengthen the management of public hospitals, and strictly prohibit doctors from violating laws and regulations. However, these measures do not nib the problem in the bud. Most importantly, while the medical reform has lowered the price of medication, it has increased the cost of diagnosis and treatment. To put it bluntly, it means "increasing the cost of consultation and reducing the cost of medication."

As a result, the increase in the cost of diagnosis and treatment from the time before the implementation of the medical reform has partially offset the falling medication costs, but in fact, the medication costs may not necessarily be reduced accordingly. This is because many doctors may not prescribe low-cost medication or basic medication that has no price differences, but expensive medication not listed in the catalog, or even imported medication. Even in many market-oriented pharmacies, cheap medication is generally placed at the corners or at the bottom of their counters, where it is difficult for consumers to find. In other words, the medical reform which changed "high medication fees" to "high consultation fees" was but a superficial change like that of packaging old wine in new bottles or changing the water but keeping the herbs in herbal medicine. This has prevented the cost of diagnosis and treatment from truly reducing.

2.1.3   Excessive Medical Treatment

A recent poll showed that over 40% of netizens felt they had previously experienced excessive medical treatment, but less than 20% chose to lodge complaints and recover their losses.

For example, when we seek medical attention, some doctors often use antibiotics for colds, and many of them prescribe the antibiotics in the form of infusions. China is not only a country with a large population but also a country with large amounts of infusions. According to survey data, the annual volume of drug infusion per capita in China is eight bottles, which is three times that of the world population.

The annual per capita consumption of antibiotics in China is about 138 grams, which is 10 times that of European and American countries. According to the statistics on abnormal deaths compiled by the Red Cross Society of China, approximately 400,000 people in China suffer abnormal deaths each year from medical harm incidents (most of them due to unsafe use of medication). This is four times the number of deaths caused by traffic accidents. Approximately 200,000 people die from the misuse of antibiotics each year.

Exploring the reasons behind this phenomenon, one will naturally think of a major drawback of China's medical system, which is the funding of hospitals through the sales of medication. In China, consultation fees are usually very low. In public hospitals, save for experienced specialists, even consultations with the Chief Physician with Professor qualification only cost a few *Yuan*. The consultation fees for ordinary doctors are even lower.

For a long time in the past, the concept of funding hospitals through the sales of medication has become a common phenomenon. Most medical representatives of large companies have strong abilities to lobby hospitals, and doctors can earn commissions off the corresponding drugs by prescribing them. Under such circumstances, selling more expensive and larger quantities of drugs equates to more revenue, and this naturally leads to excessive medical treatment. But this, in turn, aggravates the situation of "seeking medical attention being costly" for patients.

2.1.4   No Guarantee of Cover by Medical Insurance

At present, there are 570 million people insured in urban areas, and over 800 million people insured under the New Rural Cooperative Medical Scheme. Although there are some repeated enrollment and repeated statistics in the data, the total number of insured persons exceeds 95% of the total population of China. In other words, China has almost achieved universal health care.

However, as far as the reality is concerned, the actual level of protection is very low, and the medical insurance fund is on the verge of collapse, with its income falling short of its payout. For example, among the drugs in the outpatient service catalog, although 55% of the cost is reimbursable, that is to say, you can get 55 *Yuan* of reimbursement if you spend 100 *Yuan* to see the doctor for a cold, many doctors often choose drugs that are not in the catalog to earn more rebates. The costs of these drugs outside the catalog have to be borne by patients, so medical insurance becomes useless.

In addition, China's medical insurance is facing great pressure in payouts. In many places, medical costs are increasing too quickly. In some places, the costs may even double in three to five years, overwhelming the medical insurance fund, as the income falls short of the payout. According to the data in the *2014 Annual Report on China's Health Care Development*, as of 2013, the average annual increase in the income of the basic medical insurance fund for urban workers was 33.20%, while the average annual increase in payout was 34.39%. It is expected that the current income will not be able to cover the payout by 2017, and the cumulative balance of the New Rural Cooperative Medical Scheme will become negative. By 2020, the payout will exceed the funds raised in the same year by 15.38%.

2.1.5   The Tiered Diagnosis and Treatment System is Inefficient

Due to the uneven distribution of medical resources, large hospitals are often overcrowded and are always in a "wartime state." Medical disputes also occur frequently. If the patients can be effectively diverted and the primary level hospitals can be activated, then the problem can be partially resolved.

What is tiered diagnosis and treatment? Generally, it refers to separating patients according to the severity and urgency of the illness, and the difficulty of treatment. Medical institutions of different tiers will be responsible for the treatment of different illnesses.

What are the benefits of tiered diagnosis and treatment? Firstly, the diagnosis and treatment of common diseases and frequently-occurring diseases are conducted in primary-level medical institutions. The medical fee is lower, the payment threshold is lower, and the reimbursement ratio is higher. This can reduce the burden of medical expenses on patients. Secondly, for conditions that are difficult to treat or those that are complicated, patients' waiting time in large hospitals can be shortened through appointments, bed reservations and green referral channels made possible via collaborations between large hospitals and grassroots organizations. This can save patients' time and money.

Thirdly, the situation of overcrowding in large hospitals can be alleviated, so resources can be used more reasonably.

There is a popular phrase on the Internet that goes, "Dreams look fleshy, but the reality is bony." Since the implementation of the "tiered diagnosis and treatment system," which is a hot topic of the medical reform, the system has not been successful in reaching the grassroots level, and its effect is not obvious. Patients still head to where they used to visit.

According to the medical reform survey conducted in six provinces and cities by the Chinese People's Political Consultative Conference (CPPCC) National Committee for Education, Science, Culture, Health and Sports, large hospitals in the cities have been crowded with people, and the number of hospitals with more than 10,000 outpatients has increased to a large extent. On the contrary, while primary-level medical institutions had improved their staff benefits, equipment, and hardware, they still had few visitors.

The key reason behind the inefficiency of the tiered diagnosis and treatment system is that the medical conditions in primary-level hospitals do not meet the expectations of patients. Many patients who can afford to pay the fees will still choose large hospitals without hesitation because they feel that their safety is guaranteed in large hospitals compared to primary-level ones. Recently, negative news such as fatal misdiagnoses by doctors in small clinics has been reported time and again, so many patients have become more hesitant to seek medical treatment from primary-level medical institutions. Needless to say, they would be warier in cases of serious illnesses.

2.1.6 The Escalating Conflict between Doctors and Patients

On 23 March 2012, a man from Harbin with the last name Li stabbed four doctors with a fruit knife because "doctors would not attend to him." This incident resulted in one death and left three people injured.

On the morning of 25 October 2013, three doctors in the First People's Hospital of Wenling in Zhejiang Province were stabbed by a man with a dagger during an outpatient consultation. One of the doctors was killed after being stabbed several times.

At about 10 a.m. on 17 February 2014, Sun Dongtao, Director of the Department of Otolaryngology at Beiman Special Steel Staff Hospital in Fulaerji District in Qiqihar, Heilongjiang, was hit on the head with a blunt object by a man who had run in abruptly. He succumbed to his injuries and died.

Incidents like these occur every year in large and small hospitals everywhere. Conflicts between doctors and patients continue to intensify and neither party is willing to take a step back. Doctors are often nervously treading on thin ice when attending to patients daily, while patients find medical treatment cumbersome and expensive. With the addition of the patients' discomfort from their illnesses, negative emotions are always on the verge of being triggered. Such conflicts between both parties have been on the rise every year, and the situations have been reaching unprecedented levels. So much so, that some places have organized professional "medical disputes" groups.

Furthermore, according to the latest survey by *China Youth Daily*, 85.8% of doctors said that they had conducted defensive medical treatment to avoid medical risks and medical litigation, due to concerns about doctor-patient disputes. That is, the doctors try to avoid unnecessary tests and examinations, avoid receiving high-risk patients, avoid performing surgery for high-risk patients, avoid special treatments that are more challenging, as well as referrals and consultations that appear to be acts of shirking responsibility. Such defensive medical treatment prescribed by doctors will not only increase the burden on patients but may also lead to missed treatment opportunities. This is a type of soft resistance from and the inaction of the doctors' community, as doctor-patient relationships continue to deteriorate. This is also the negative effect that is the most difficult to eliminate.

In this tense doctor-patient relationship, it is difficult to conclude who is in the right or wrong, because in reality, once a dispute occurs, it is unavoidable for both the doctor and the patient to be seriously harmed.

On 6 February 2015, Jiang Yu, Associate Researcher at the Development Research Center of the State Council, published the article *Do not make 'subversive mistakes' in medical reform*. In it, he mentioned that there are risks of making "subversive mistakes" in the medical reform. They are mainly manifested in the marketization, commercialization, and privatization of medical health care, which is a diversion from public welfare, fairness, and universal access to basic medical and health care services.

For many years, there have been disputes over the two paths of medical reform – public welfare and marketization. In 2009, the central government decided on the direction of government leadership and public welfare, but its implementation was far from sufficient. The supply of basic medical and health care services was still weak, and basic medical insurance cannot alleviate the burden of medical expenses from patients sufficiently. Public hospitals are still operating under the old strategy of funding hospitals through the sales of medication, reaping profits in the name of a public entity. The cost of medication has been on a rapid rise, which has negated the effects of medical reform at the primary-level and basic medical insurance. The burden of medical expenses for the masses has not been alleviated.

Medical reform has made the problems brought about by the traditional medical system ever more prominent. This can be analyzed from the following aspects:

Traditional medicine has almost come to an end. If reform is not carried out, doctor-patient relationships will inevitably be escalated. Therefore, in my opinion, the way out for traditional medical treatment is to actively use the Internet, ride on the wave of "Internet Plus" health care, and shake up the entrenched traditional medical system.

First, the top-level design should be completed on the national level. Second, hospitals themselves should be prepared for reform. Third, the market should be prepared to build connections. Fourth, people in the industrial chain need to change their mindsets and welcome changes. Fifth, there will be a change in people with vested interests. Sixth, using the Internet, big data, artificial intelligence, cloud platforms, wearable medical devices, and other products from the new technology era would help to re-establish a new medical ecological chain that truly places patients at the center.

If the top-level design of the "Internet Plus" medical reform is done at the national level, but the relevant parties in the industrial chain do not change accordingly and seize this opportunity, there will inevitably be a disruption.

2.2　Continuous Informatization of Medical Health Care

With the emergence of the Internet and the mobile Internet era, the rigorous and prudent medical industry has finally been swept by this strong wind and disrupted. In March 2015, the General Office of the State Council issued the *Outline for the Planning of the National Medical and Health Service System (2015–2020),* which mentioned for the first time that China will launch plans for a healthy China cloud service. This could stimulate the rapid development of the Internet medical industry. It would also boldly explore and actively innovate areas like transforming the management system of public hospitals, strengthening medical insurance payment and its monitoring role, building a service system for the coordinated development of various medical institutions and accelerating the informatization of medical health care.

At the 11th meeting of the Central Comprehensively Deepening Reforms Commission held on 1 April 2015, the *Guidance on the Pilot for Comprehensive Reform of Public Hospitals in Cities* was reviewed and approved. It was emphasized during the meeting that the basic positioning of public hospitals as a place for public welfare must be adhered to, and the profit-seeking mechanism of public hospitals must be abolished.

In the next five years, China will launch plans for a healthy China cloud service and actively utilize new technologies such as mobile Internet, the wider Internet of Things, cloud computing, and wearable devices. This will be done to promote health care informatization and smart medical services that benefit all. As well as to promote the application of big data in health care, gradually change service models, and improve services and management capabilities.

It is predicted that the market for informatization of China's medical and healthcare industry has maintained rapid development since 2012. The informatization of county hospitals and community health care institutions, the construction of regional health informatization platforms, and the construction of public health systems in various provinces have become the main driving force for the informatization of the medical industry.

Mobile application systems have begun to be used in large hospitals, social media sites continue to try providing new medical service models, while cloud computing technology has been vigorously promoted and gradually implemented. Statistics show that the scale of the informatization market for China's medical and healthcare industry in 2012 was about 17 billion *Yuan,* an increase of 21% from 2011. By 2014, this market size was expected to reach or even exceed 20 billion *Yuan.* In 2015, the medical informatization market was expected to exceed 30 billion *Yuan,* and the market growth rate was expected to exceed 15% in the next three years.

The reform of public hospitals will continue to deepen, driving the upgrade of information systems. The implementation of the separation of prescribing and dispensing will begin from Beijing and Shanghai and extend to other provinces. The number of hospitals implementing such separation will increase. With the implementation of the separation of prescribing and dispensing, the pricing, calculation, and supervision of medical services will now be on the agenda. The separation of prescribing and dispensing has driven the upgrade of medical information systems such as the HIS system, boosting the growth of the medical software and its service market. At the same time, public health care services and management have been strengthened, and investment in information systems for public health care management has increased.

2.3   Current State of Medical Investment in China

2.3.1   The Potential in China's Medical Industry

Relevant data shows that medical and health care expenditure has become the third-largest consumer expenditure after food and education in China, but the overall size of China's medical and health care industry accounts for less than 2% of China's GDP. The number of visits made for medical diagnosis and treatment in China has increased by more than two times since 2000. In recent years, the growth rate has been maintained at around 8% to 10%. The government has clearly stated its intention to increase the market size of the medical service industry from 2 trillion to 8 trillion, which equates to a growth of more than 20% year on year. Hence, the industry has huge room for growth.

Moreover, domestic medical expenditure has accounted for less than 5% of China's GDP over a long period, while the medical expenditure in the U.S. has accounted for up to 16% of its GDP. Even compared to countries whose GDP per capita is comparable to China, China's medical expenditure still has a lot of room for growth.

According to statistics, the size of China's health care industry was close to 2 trillion *Yuan* in 2013. The overall scale is expected to exceed 3 trillion *Yuan* by 2017. China's health care industry will be ushering in rapid development.

In the field of mobile health care, data from iiMedia Research shows that China's market size amounted to about 2.8 billion *Yuan* in 2014, and it will exceed 12.5 billion *Yuan* by 2017, with a compound annual growth rate of over 70%. The website abaogao.com is more optimistic. Its data shows that the size of China's mobile health care market will amount to about 15 billion *Yuan* by 2017, of which the market size of wearable medical devices will be worth approximately 4.77 billion *Yuan*.

2.3.2   New and Old Capitalists Begin to Set up and Expand the Boundaries of their Territories

Global Internet healthcare has attracted approximately USD 8 billion in venture capital over the past four years, and some innovative companies have developed rapidly and had been successful in their IPOs. They include large-scale financing with transaction volume reaching billions of dollars, such as Veeva Systems and Castlight Health. These companies have made this industry one of the hotspots for venture capitalists and acquirers.

According to data from Analysys, capitalists began watching the mobile medical industry in 2010, and subsequently, large numbers of capitalists began entering the market in 2012. With the maturity of foreign mobile medical industry business models and the rising enthusiasm for capital investment, the mobile medical investment in China also entered a period of high enthusiasm in 2014 (Figure 2-1).

Besides, it can be observed from the amount of investment in the mobile medical industry that it has increased year on year (Figure 2-2).

2010H1-2014H1 Investment count in China's mobile medical industry

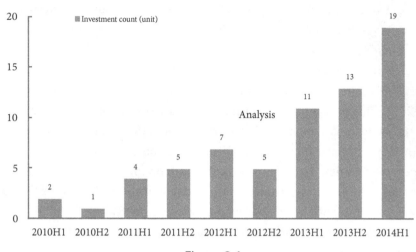

Figure 2-1

2010H1-2014H1 Amount of investment in China's mobile medical industry

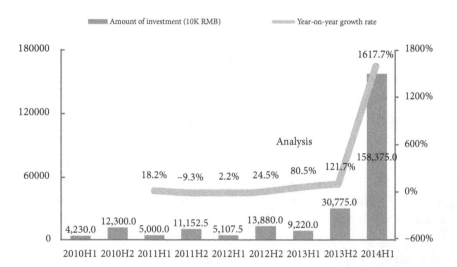

Figure 2-2

In the first quarter of 2015, 25 Internet medical companies received investment, with a total financing amount of USD 408 million, an increase of 78.5% from the previous month and an increase of 38.7% year-on-year (according to data from VCBeat). For details, please refer to Table 2-1.

Table 2-1

Project Name	Financing Details	Primary Business
DXY.cn	In March 2010, it received the first round of investment worth USD 2 million from DCM In end 2012, it received a series B financing of USD 10 million led by Shunwei Capital and followed by DCM In September 2014, it received 70 million *Yuan* worth of investment from Tencent	Online healthcare
chunyuyisheng.com	In 2011, it received an angel investment of some USD 3 million from BlueRun Ventures In September 2012, it received a series B financing from BlueRun Ventures In August 2014, it completed a Series C financing worth USD 50 million	Mobile Internet company with mobile health care as its business direction
91huayi.com	In September 2010, it received 25.2 million *Yuan* worth of investment from Silicon Paradise and Jiangsu GOVTOR In March 2010, it received its first round of investment worth 30 million *Yuan* from Sequoia Capital	Education for the medical profession: chronic disease management, health education
OmeSoft	In March 2011, OmeSoft received RMB 2 million worth of investment from Shanda	Developer for public health and medical mobile applications
Andon Health	In September 2014, it received USD 25 million worth of investment from Xiaomi	Medical and healthcare electronics provider

(continued on next page)

31

Project Name	Financing Details	Primary Business
codoon.com	In July 2011, it received 22 million *Yuan* worth of series A investment from Shanda In November 2014, it received 60 million *Yuan* worth of series B financing from Shenzhen Capital Group	Smart wristbands
dayima.com	In April 2013, it received several hundreds of USD worth of investment from Bertelsmann Asia Investments and ZhenFund In June 2012, it received millions *Yuan* of investment from ZhenFund, Tisiwi and others	Mobile phone application for women's menstrual care
Hinacom	In December 2011, it received its first round of investment worth 30 million *Yuan* from Sequoia Capital	Provides doctors with medical imaging information systems and services
guahao.com	In December 2010, it received tens of millions of USD from FengHe Group and Cybernaut in its series A financing In January 2012, guahao.com conducted a series B financing without revealing the specific amount of investment In October 2014, it received USD 100 million from Tencent in its series C financing	Health consultation and medical guidance platform
91160.com	In May 2013, it received several million *Yuan* of angel investment from HGI FINAVES China Fund In January 2014, it received a joint investment of 10 million *Yuan* from QF Capital, Beijing Upring Investment, and others in its series A financing In May 2015, it received nearly 100 million *Yuan* of investment from Co-Stone Capital	Online platform for appointment registration

(continued on next page)

Project Name	Financing Details	Primary Business
Distinct HealthCare	In January 2014, it received tens of millions of RMB worth of investment from Matrix Partners China in its series A financing In April 2015, it received USD 17.5 million in its series B financing	A high-quality private medical institution
APPSCOMM	In May 2014, it received 60 million *Yuan* of investment from China Eagle Electronic Technology in its series A financing In October 2014, it received an investment of USD 10 million from Intel Capital in its series B financing	Smart wearable devices core technology and cloud development and application
Oranger	In November 2014, it received an investment of USD 5 million from HighLight Capital in its series A investment	Medical wearable equipment company, with its main product being a snoring monitor

2.4 BAT Makes Plans for "Internet Plus" Medical Care

2.4.1 Alibaba's Grand Health Strategy

"From now on, Alibaba wants to be involved in the two industries of health and happiness. How to make people healthier and happier? It is not done by building more hospitals and finding more doctors, nor is it done by building more pharmaceutical factories. If we (invest) do it right, in 30 years, doctors would become unemployed, and there would be fewer hospitals and a lot fewer pharmaceutical factories. Those will signal that we have done it right." It is evident from this speech delivered by Jack Ma that Alibaba's future strategy will focus on the medical industry.

In January 2014, Alibaba and YF Capital jointly spent 1.037 billion to acquire 54.3% of CITIC 21CN's shares. Shortly afterward, CITIC 21CN issued a notice to announce that it had officially been renamed as "AliHealth." Alibaba had thus officially announced its entry to the medical industry.

33

In May 2014, Alipay announced the "Future Hospital Plan," declaring that Alipay would open up its platform capabilities to medical institutions, which include its accounting system, mobile platform, payment, and financial solutions, cloud computing capabilities, and big data platforms. In September that same year, Alibaba also launched the Pharmacy O2O project by working with pharmacy chains in Hubei, Guangdong, Shanghai, Chongqing, and other regions. It seized the market for pharmacy O2O by taking advantage of its Internet payment capabilities and strengthening the pharmaceutical e-commerce link.

In February 2015, the AliHealth Cloud Hospital platform was launched, and it was positioned as "a network platform that integrates the entire medical system and the entire chain of resources, and provides comprehensive medical services." In April, the AliHealth Cloud Hospital platform website was officially launched.

Since the second half of 2014, Alipay's Future Hospital Project, AliHealth Cloud Hospital (called "Yidiegu"), Cloud Hospital (Alibaba Cloud), and other ideas revolving around Alibaba's grand health care blueprint have begun to materialize one after another.

Alipay's "Future Hospital"

In May 2014, Alipay officially announced its "Future Hospital" strategic plan, whereby Alipay would open up its platform capabilities to medical institutions in the future. These include its accounting system, mobile platform, payment, and financial solutions, cloud computing capabilities, and big data platforms. This was done to help hospitals become more efficient in the mobile Internet era.

The medical treatment process would generally be:

Online registration, waiting for consultation—offline consultation—online inspection and payment—offline inspection—online report collection—offline diagnosis—online payment for medication—offline collection of medication for treatment (Figure 2-3).

In the hospital consultation segment, Alipay would also integrate the indoor navigation technology of Gaode Maps and provide patients with technical support including free Wi-Fi connection and indoor navigation in the hospital, to guide users to complete the entire process of consultation, inspection, payment, and collection of medication in the shortest time possible. At present, Alibaba has attained cooperation intentions with nearly 50 Tier 3A hospitals in major cities across the country. In 2015, more than 100 hospitals are expected to join the "Future Hospital" plan.

Medical treatment process

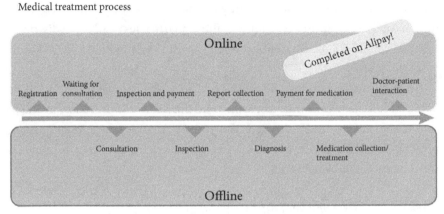

Figure 2-3 Alipay's medical treatment process

In addition, in the era of universal medical insurance, Alipay's "Future Hospital" will also integrate the function of the real-time settlement of medical insurance. In the future, users will only need to pay the portion of the costs they are supposed to bear using Alipay. Alipay will make a real-time claim for the portion claimable under medical insurance.

Seizing the pharmacy O2O market first

Promoting true "separation of prescribing and dispensing" is an important part of the medical dream that Ma Yun wants to create.

On 25 June 2014, Alipay announced a partnership with ChinaSoft International, to jointly test the waters of the mobile payment business in the pharmaceutical industry. The main objective of the partnership was to make way for Alipay in the field of pharmacy payment, to provide users with mobile payment service when they are purchasing medication.

Alipay revealed that it will incorporate pharmacy services into its entire "Future Hospital" plan in due course. It will also launch services such as the distribution of medication nearby and regular delivery of medication for chronic diseases, following the promotion of medical reform policies such as the separation of prescribing and dispensing, and electronic prescriptions. Furthermore, offline pharmacies and yao.tmall.com will be connected in the future. After users place orders online, offline pharmacies nearby can distribute the medication. All pharmaceutical retail companies can join Alipay's "Future Hospital" plan.

In the future, patients will only need to seek medical attention at an organization that works with AliHealth, and the electronic prescription will be sent to the patient's AliHealth app. The patient can open the app and use the prescription to place an order with a pharmacy nearby, then wait at home for the medication to be delivered.

Alternative supervision with the "Medicine Safety Plan"

In July 2014, based on relevant data from CITIC 21CN after the acquisition, Alibaba announced the launch of the "Medicine Safety Plan," which is equivalent to supervision under the free market system. This is essentially the same service that Taobao is currently trying to provide.

Each box of medicine has a unique medicine supervision code. Users just need to scan the barcode and medicine supervision code on the packaging of any box of medicine in China using the mobile Taobao or Alipay app and they will be able to obtain information such as the authenticity of the medicine, usage instructions, things to avoid, its production batch and circulation process (Figure 2-4). This will alleviate consumers' concerns about the authenticity and safety of medicines to some extent.

One of Alibaba's objectives in rolling out the medicine scanning service is obviously to pave the way for e-commerce in medicine, but there is more to Alibaba's medical "ambition." According to its plan, Alibaba will create a comprehensive medical data service platform, where data such as medicine

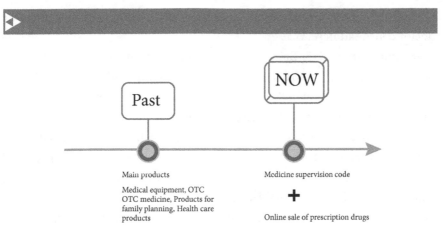

Figure 2-4

circulation, functions, consumer purchases, and medicine usage behaviors will be integrated to form a health database. Building on this foundation, pharmaceutical companies can manufacture and store medicine according to the consumer purchase situation, patients can approach manufacturers to customize medicine, and regulatory bodies can implement tracking verification on every bottle of medicine.

At the core of Alibaba's "Future Hospital" is the control of the flow of funds in the entire process of a patient's medical consultation, as well as the flow of medicine consumption. The "Future Hospital" plan can build on this and dig deeper to derive commercial value and transform the structure of the current ecology.

2.4.2   Baidu's "Cloud" Doctor Plans

Baidu appears to be more cautious and calm than Alibaba and Tencent in planning its steps in the medical field. In 2013, "Baidu Health" was launched. In June 2013, Baidu's pre-medical smart consultation platform was also launched. In December 2013, Baidu established dulife, a smart portable device brand. The following year, dulife smart wearable products were released. In July 2014, "Beijing Health Cloud" was officially released, becoming the first smart health management project in the world. In October 2014, Baidu launched an online doctor consultation service.

Stepping into 2015, Baidu made even more frequent, big moves in the medical field. First, it struck a strategic partnership with 301 Hospital, an authoritative institution in the medical field in China, to jointly build a mobile Internet medical online platform. Next, Baidu used the "Baidu Doctor" app as a connection point to closely link up patients with medical services. In addition, Baidu also strategically invested in the health care network this year, to extend Baidu's medical services offline. Recently, it formally achieved strategic cooperation with the National Health and Family Planning Commission and became the latter's official platform for the promotion of new medical reform. Baidu has again been endorsed by the national authority, which showed that it has an exceptionally powerful influence.

Baidu's biggest advantage lies in its big data. As a search engine, Baidu has accumulated a large amount of data. On this basis, there is much room for imagination on the scenarios for the application of the data. Examples include monitoring of epidemic situations, disease prevention and control, clinical research, medical diagnosis decision-making, medical resource management, family telemedicine, and various other aspects. What Baidu wants to achieve in the medical industry is to truly "connect people and services" with the Internet.

In planning its next steps, Baidu mainly focuses on the current "pain points" experienced by patients during medical consultations, which is the present big issue in China, where "it is difficult to see a doctor." Before "paying for medical services," a patient has to spend a lot of time studying his or her illness and deciding on a suitable hospital, department, and doctor.

In response to this phenomenon, Baidu President Zhang Yaqin pointed out that the biggest problem in the medical industry lies in the high degree of asymmetry of medical resources between regions and between different hospitals. He said, "the problem that Baidu Doctor wishes to solve is to allow users to find the best doctor nearby as quickly as possible, then use Baidu's positioning technology, search technology, and machine precision matching technology to allow everyone to find the doctor that suits them best."

Zhang Yaqin believes that Baidu's high traffic can bring users, who are patients, to the hospitals. Hospitals can provide a wealth of hospital and doctor information, registration, and other outpatient resources for Baidu's doctor-patient dual-selection platform, to create a closed-loop with medical services.

The next step is to close the loop online and offline. The online service of Baidu's doctor-patient dual-selection platform can eventually provide users with a better medical consultation experience and open up online and offline services through the hospital's offline service team.

The third point is to achieve a closed-loop for the product itself. That is the closed-loop of doctor appointments, registration, and payment.

Zhang Yaqin pointed out that mobile health care will be the trend for future development. In the future, besides seeing a doctor, one can find a doctor, book an appointment with the doctor and leave a rating for the doctor online. This will simplify the entire medical treatment process.

Baidu will integrate its technology accumulated over the years, including cloud platforms, big data, payment, positioning, and other technologies to provide support for and to transform major hospitals and use the hospital as an entry point to create a closed-loop for doctors, medicine, patients and other links.

2.4.3 Tencent's "Smart Health Care" Takes Off

At the two sessions meeting in 2015, Ma Huateng put together a proposal to use social forces to optimize online registration services and establish a national-level "Chinese brain" plan. In terms of medical treatment, Ma Huateng specifically recommended that the government connects the Internet to the current medical system, comprehensively improves the informatization level of hospitals, lift restrictions on online registration in some areas and increase the proportion of hospitals implementing online registration year by year. All of these steps will help to solve people's concerns that seeing a doctor is expensive and difficult.

At the end of 2014, Tencent heavily invested USD 170 million and took over the ownership of DXY.cn and guahao.com, quickly executing its plans for the Internet medical industry (Table 2-2). Jiang Haoran, Director of Strategic Development at Tencent, said that its deployment in Internet health care is mainly done to solve the four major pain points: poor medical experience in large and small hospitals, poor communication between doctors and patients, cost of medical treatment, and the lack of personal health and medical data.

Table 2-2   Investment details

Investment organization	Company	Funding series	Amount	Time
Tencent	Tencent Industrial Fund Co., Ltd.	guahao.com	Series C	USD 100 million	October 2014
	Tencent Industrial Fund Co., Ltd.	DXY.cn	Series C	USD 70 million	September 2014
	Tencent Industrial Fund Co., Ltd.	Binkepurui	Series B	USD 21 million	June 2014
	Jingdong		Angel investment	Millions of RMB	
	Decent Capital	linjiayisheng. com			January 2014

Tencent has always been a big player in the social sector. When it entered the health care industry, it leveraged on this advantage and took its first step in smart health care – WeChat registration. So far, more than 1200 hospitals across the country have bought into Tencent's "smart health care" solution, which uses WeChat as the service platform. On top of that, nearly 100 hospitals use WeChat for the whole process of medical consultation, more than 120 hospitals support WeChat registration, and over 3 million patients have benefited from these services.

The WeChat Smart Hospital, launched in 2014, builds on the foundation of "official accounts + WeChat pay," combining WeChat's mobile e-commerce portal, as well as user identity verification, data analysis, payment settlement, customer relationship maintenance, after-sales service, rights protection, and other segments. The result is an improvement in the connectivity between doctors, hospitals, patients, and medical equipment, which simplifies the entire medical treatment process.

On 7 December 2014, the QQ Health Center was launched. After connecting to the QQ Health Center on the mobile phone through wearable devices, users can view their health statuses in real-time through an open health management social platform.

Features of the WeChat Smart Hospital include WeChat registration (Figure 2-5), reminder while waiting, payment of consultation fees, access to various electronic reports, consultation navigation, and push notifications for suspension of consultation by specialists. Besides, WeChat can deliver electronic reports in real-time, as well as reminders from the doctor after the user has left the hospital. Looking at the entire process, Tencent's WeChat Hospital includes a full range of services before, during, and after medical treatment. Particularly for the process during treatment in the hospital, the use of WeChat saves a lot of time.

To take a further step in the Internet medical industry, Tencent also began putting in the effort on medical hardware. On 20 January 2015, Tencent launched a smart blood glucose meter called Tang Daifu (Figure 2-6). This smart blood glucose meter is equipped with a 4.0-inch IPS high-definition display has a design exactly like a smartphone, it supports Internet connection, and can automatically record and make measurement results accessible via WeChat. It is designed and developed by Tencent's WeChat team, while the blood glucose monitoring data is managed by WeChat. When the patient's level of glucose is abnormal, the meter will also send warning messages to the patient and his or her family members.

Tencent's blood glucose meter faces a similar problem with all other medical hardware in the market currently, that is, it asks how it can ensure the accuracy of its inspection data. But in any case, this was just one of Tencent's attempts to test the waters. The rise of the entire Internet health care industry in the future will be closely linked to medical smart hardware.

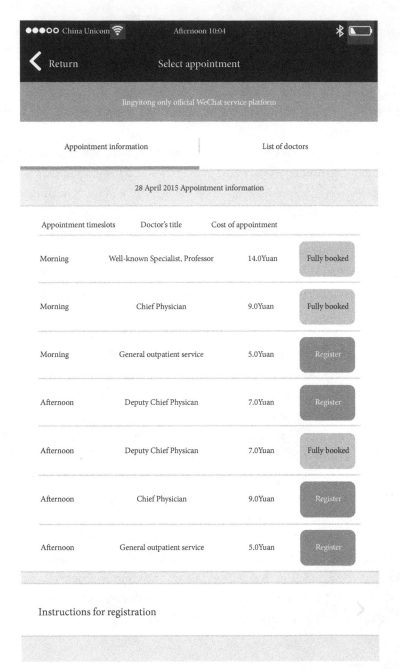

Figure 2-5

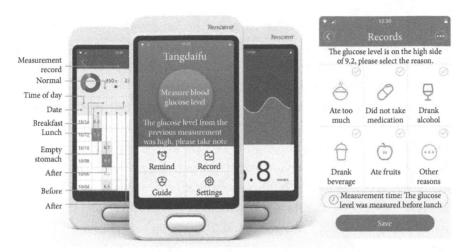

Figure 2-6

2.5 Registration and Payment have Become the Entry Point for "Internet Plus" Health Care

Whether it is Alibaba or Tencent, the easiest way to enter the health care industry is through the "lightest" segment of medical treatment, which is appointment making and registration. Judging from Tencent's investments made in the medical guidance platform guahao.com, the functions of appointment registration, payment, queuing for consultation, and inquiry about inspection reports have been made available in more than 900 key hospitals in the 23 provinces across the country. The distribution of red packets for registrations made on "Weiyi" (guahao.com) has also been quietly launched. In Guangzhou, users only need to follow the service account "Guangzhou Jiankangtong" on WeChat, and they will be able to gain access to services such as appointment registration, medical fee payment, and health records management in 60 hospitals in the city. At the same time, the "Pay (WeChat Pay) a penny" registration function will also soon be launched.

On the surface, the competition between Tencent and Alibaba in the medical field is likened to the fight in hailing a cab. It appears that they are both competing in the area of medical fee payment. Yet, payment is only an entry point. Both of them have set their eyes on the huge medical and health market that follows it. Looking at the current situation, Tencent's planned moves are different from Alibaba and Baidu.

At the end of October 2014, guahao.com, a platform that provides medical guidance and health-related queries for the public, received an investment of USD 106.5 million led by Tencent. The website guahao.com had 100,000 registered doctors and 37 million users registered using verified true identities and it was connected to the information systems of more than 900 hospitals in the 23 provinces across the country. Tencent's "smart medical" plan is more focused on gaining entry to the medical industry through doctors. It aims to connect medical professionals, hospitals, and users, to build a "smart hospital."

The development of this plan will involve doctor-patient communications, links to medical insurance, indoor navigation, referrals through different levels, direct payment settlement for commercial insurance, and medication purchase and payment. Jingdong will particularly be tasked to complete the medical purchase and payment segment. This is also one of Tencent's strongest advantages. The above segments are those which Tencent must optimize next.

Unlike Tencent, Alibaba's plan for the Internet health care industry not only involves opening up all medical fee payment channels using Alipay but also seizing the medicine market. With the introduction of policies such as the trial of third-party sales platforms online for medicine and the online sale of prescription drugs, the pharmaceutical online sales industry has become more mature, and many online big players are creating pharmaceutical e-commerce platforms. For Alibaba, the veteran, this is exactly where its huge advantage lies.

In China, a market segmentation war has been launched in the Internet health care industry, but the big players who are eyeing this industry are not the Internet big players such as Tencent, Alibaba, and Baidu. Xiaomi's Lei Jun and Jingdong's Liu Qiangdong will not sit back and watch BAT divide the domestic mobile medical and health care industry among themselves. Moreover, in February 2015, Suning.com, GOME, and Dangdang had also declared that they wanted a share of the Internet health care market each. They will get it by entering the market through the remote mobile medical treatment.

CHAPTER 3

Current Status of Wearable Health Care

From the initial Google Glass to Apple Watch in 2015, the wearable device industry has been continuing to up its game, offering more interesting experiences for users with surprising new functions and effects. This has also opened the public's eyes to new possibilities.

Evidently, wearable devices today have become very popular in tech circles. Whether it is the Internet or smart homes, as far as smart devices are concerned, wearable devices will always be involved. This is because wearable devices are the key to everything. Without them, there will be no medium for everything else. They are akin to a person losing their heart.

Particularly in the medical field, wearable devices have been regarded as the best entry point for hardware, because the devices fully trigger users' desire and motivation for a healthy body, and they are capable of meeting such rigid needs. Wearable devices assist users in managing their health and developing good lifestyle habits. As such, wearable devices are bound to bring about a disruptive revolution to the entire health management and medical industry. Indeed, this market is developing at an accelerated rate in China (Figure 3-1).

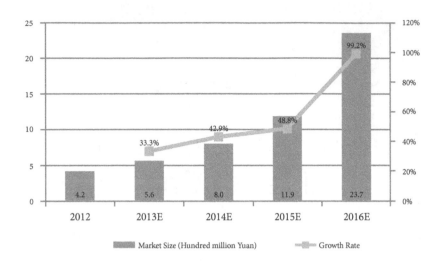

Figure 3-1 2012-2016 Forecast of China's wearable mobile medical device market size

The data shows that in 2012, China's wearable mobile medical device market sales amounted to 420 million *Yuan,* and the market size is expected to exceed 1 billion *Yuan* in 2015.

3.1 Concept

So, what are wearable medical devices? How are they related to the current wind draft of the Internet Plus health care? First, let us take a look at an example:

Stanford University developed a new wearable sensor that can perform tests on the heart. This sensor is paper-thin and is only as large as a post stamp. Heart disease patients just need to attach it to their wrists and it will be able to perform heart tests anytime and anywhere, without affecting their daily work and life. Mobile internet research media believes that once the accuracy of this device is standardized, it will be able to replace the current method of measuring blood pressure, which requires inserting a catheter into the blood vessel.

From this example, we have seen that wearable sensors have replaced medical devices that originally needed to enter the human body. This reduces the physical pain experienced by the patient during the inspection process. Furthermore, in the future, this inspection method through sensors will become more accurate than traditional methods.

In other words, wearable health care is equivalent to sensor health care, supplemented by wireless communication, multimedia, and other technologies, and placed across various parts of the human body. Sensors can detect various body signals in real-time to prevent diseases. Once the user falls sick and requires medical treatment, the device can automatically assist the user in completing the treatment process and sending back a report to the doctor. It can even immediately deliver medication to the user. This whole process may be completed without the patient realizing that he or she is even sick. Finally, after receiving medical treatment, the device can arrange dietary and rest plans according to the doctor's advice, to optimize the effect of the medical treatment and medication consumption.

All in all, wearable health care is a complete system and a huge platform. If any link is missing, it will be difficult to maximize the value of wearable devices in the medical industry. Many of the present wearable devices for daily use such as watches, wristbands, apparel, and footwear can measure various vital signs of its users, but they are still in the early stages of health management. They are far from truly being utilized in the entire wearable medical industry. For wearable devices, the medical industry is where it can maximize its value. Health management is just a baby step.

Wearable health care is part of internet health care, and is, at its core so, because only wearable devices can fully support internet health care. Internet health care would be a baseless and impractical concept without wearable devices. Specifically, wearable devices are indispensable for electronic medical records, remote diagnosis and sample data collection for research.

The biggest advantage of wearable devices lies in their ability to store personal health information for future analysis anytime, anywhere. For individuals, the future trend would be one where everyone has his or her device and serial number, just like an ID card, which can be used for managing personal health.

3.2   Favored by PE/VC

It is understood that smart wristbands stand out the most among wearable devices. Canalys, a market research company, estimates that global shipments of wristbands will increase by 129% from 2014 to 2015, amounting to 43.2 million units, and shipments of smart wristbands and smartwatches that can run third-party applications, like Apple Watch, will amount to 28.2 million units (Figure 3-2). In 2017, this figure is expected to increase to 45 million.

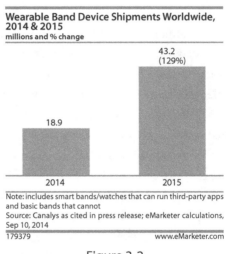

Figure 3-2

Jawbone, the big player in the smart wristband industry, currently has a market value of about USD 3 billion (Figure 3-3), and more than USD 700 million in its total amount of financing. In early 2015, Jawbone completed another round of financing worth USD 300 million in addition to its original amount. Its investor was the famous BlackRock. The company, headquartered in San Francisco, the United States, has three iconic products—Bluetooth headsets, speakers, and UP smart wristbands.

In addition, Fitbit, a wearable device startup that focuses on smart

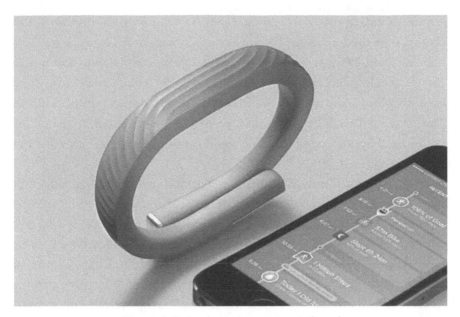

Figure 3-3 Jawbone's smart wristband

wristbands, applied for listing in May 2015, intending to raise USD 1 million. The investment banks that raised funds for them include Morgan Stanley, Deutsche Bank and Merrill Lynch. This 8-year-old company is likely to become the first professional wearable device company that is publicly listed.

In the prospectus, Fitbit said that its user base is about 95 million large, and this includes members with annual cards worth $50, wearable device users, Fitbit account holders, and individuals who had uploaded their personal information onto the Fitbit data center. In March 2015, the personal trainer app released by Fitbit already had 2 million users on the iOS platform.

For a company that specializes in smart wristbands, such an achievement invigorated the entire smart hardware industry.

The website, iiMedia.cn, analyzed this situation, and believes that as the wearable device industry gains popularity, wearable devices will also achieve rapid development in the medical industry. As people pay more attention to their health, the market size of wearable mobile medical devices will continue to expand.

Survey data from ABI shows that about 30 million wireless wearable health sensors were used in the field of medical electronics in 2012, which is a 37% increase from 2011. At the same time, the wearable mobile medical device market sales in China are expected to amount to 420 million *Yuan* in 2012 and exceed 1 billion *Yuan* in 2015. By 2017, the scale is expected to reach 5 billion *Yuan*. ABI predicts that applications for remote patient monitoring and online professional health care will account for 20% of the overall wearable wireless device market in 2017.

Mobile internet, wearable devices, big data, cloud computing, and other new technologies are constantly changing our understanding of medical treatment. Traditional health care mechanisms and hospital business models will gradually be replaced by more convenient and smart mechanisms. Because of this general trend, whether in China or other countries, a large amount of capital will be directed towards the new medical industry. In particular, the market for medical software and hardware that leverage wearable devices will become one of the next investment areas.

There are two main types of companies that are currently working on wearable health care, and the first is startup companies. They usually ride on the popularity boosted by increasing exposure to attract the attention and investment of venture capitalists. Their choice of medical devices is entertainment-grade and mildly smart devices. Some of these companies turn out successful, but most of them will find it difficult to sustain their business. Another type is traditional enterprises, which mainly focus on the industrialization of wearable devices and act according to their future goals. Their consideration will be profitability. Hence, these enterprises will choose professional-grade medical devices.

According to GSM's measurement standards for the mobile medical industry, medical device manufacturers and content and application providers account for about 39.83%. It is predicted that the scale of China's wearable and portable medical device market will be close to 5 billion *Yuan* in 2017 (Figure 3-4).

Although the wearable medical industry has promising development prospects, its growth will ultimately rely on capital and technology. In 2014, the following six categories of enterprises have gained more investments: internet medical treatment, big data analysis, consumer participation, digitalized medical devices, telemedicine, personalized medical treatment, and health management.

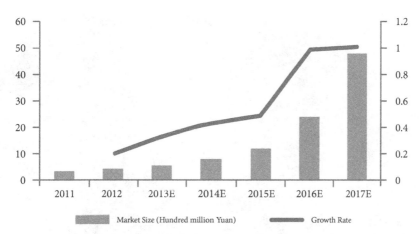

Figure 3-4 The scale of China's wearable and portable medical device market

In recent years, the following domestic and foreign investment and financing projects in the field of wearable health care had a relatively greater impact:

Kang Blood Pressure Monitor

Kang blood pressure monitor (Figure 3-5) is a hypertension monitoring and guidance tool that combines software and hardware. It works by using a blood pressure monitor hardware worn on the upper arm to read blood pressure and heart rate data. After comparing the data with the cloud, the blood pressure monitor provides health guidance such as exercise, diet, medication, and medical treatment to the user via an app.

In 2013, the Kang blood pressure monitor received RMB 4 million worth of angel investment. In early 2014, it completed another series of pre-A financing worth RMB 30 million. The direction for this round of financing was relatively clear. About 60% to 70% would be invested in product development that year and the next, while the remaining amount would be used as market expenses for educating users and growing the user base.

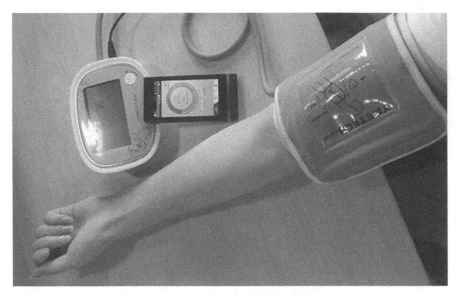

Figure 3-5 Kang Blood Pressure Monitor

Oranger Snoring Monitor

In March 2014, Oranger released a snoring monitor for medical treatment, which is mainly targeted at the initial screening of the sleep apnea syndrome. In addition, Oranger jointly established a community screening center for sleep apnea syndrome with Tier 3A hospitals to lay out its plans for healthy communities, to realize the "private doctor" service model.

At the end of 2014, Oranger received an investment of USD 5 million from HighLight Capital in its series A investment. This round of financing would be used to strengthen the accumulation of technology and at the same time lay down the plans for O2O healthy communities. It will also provide users with a one-stop health solution and a more comfortable user experience.

Huami Corporation's Xiaomi Mi Band

Huami Corporation, founded in January 2014, is a company that specializes in smart hardware and services including smart wearable devices, identification, and human body data. In the early stages of the company's establishment, it had obtained tens of millions of RMB from Xiaomi Corporation and Shunwei Capital in its series A investment.

In Xiaomi's ecological chain, Huami Corporation is the only company engaged in wearable devices. Its main product is the Xiaomi Mi Band (Figure 3-6). In December 2014, Huami Corporation announced that it had received USD 35 million worth of series B financing, led by Gaorong Capital, and followed by Sequoia Capital, Morningside Venture Capital, and Shunwei Capital. The valuation of the company after financing was over USD 300 million.

Statistics show that as of the end of November 2014, the Xiaomi Mi Band had already shipped 1 million units in three months, owing to its strong marketing capabilities and extremely affordable price of 79 *Yuan*. It has become a wearable device with the largest market share.

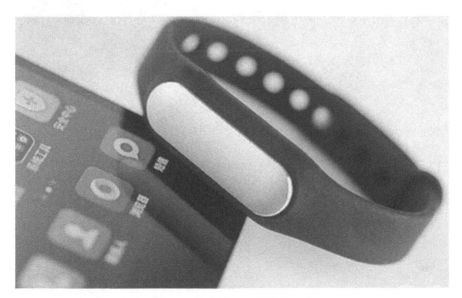

Figure 3-6 Xiaomi Mi Band

Chengdu Codoon Smart Band

The Codoon Smart Band (Figure 3-7), a product of Chengdu Codoon Information Technology Co., Ltd can be combined with mobile phone apps to monitor users' exercise and sleep, and establish a personal health profile on their cloud.

In April 2011, Codoon, the data platform service provider for wearable devices, first received an angel investment of RMB 22 million from Shanda. In March 2014, Codoon obtained RMB 60 million from Shenzhen Capital Group and CITIC Capital in its series A investment. In November 2014, it received USD 30 million in its series B financing, with a joint investment made by SIG and SB China Capital.

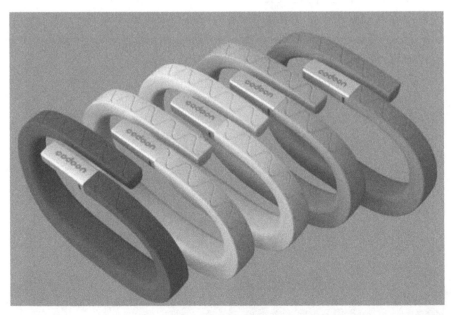

Figure 3-7 Codoon Smart Band

Appscomm Smart Watch

Founded in May 2011, Appscomm is an innovative technology enterprise that specializes in wearable smart core technology and cloud development and application. Its smartwatch business spans three major areas including smart aging, health management, and fashion sports, which provide terminal services for those services. Its products include Fashioncomm smartwatches and Bluetooth wristbands.

In May, 2014, Appscomm obtained RMB 60 million from China Eagle Electronic in its series A investment. In October that same year, it obtained USD 10 million from Intel Capital in its series B financing.

Andon Health Blood Pressure and Blood Glucose Level Measurement Instruments

The main research areas of iHealth, a subsidiary of Andon Health, are in medical platforms and products based on big data of health, as well as blood pressure and blood glucose measurement products that solely measure blood pressure and blood glucose levels, respectively.

In 2014, iHealth announced that it received USD 25 million worth of investment from Xiaomi in its series A investment. In the future, iHealth and Xiaomi will engage in further cooperation in user experience, Xiaomi e-commerce, and cloud services, to jointly create a mobile health cloud

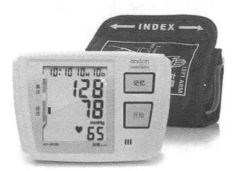

Figure 3-8 Andon Health Electronic Blood Pressure Monitor

platform. This will successfully fill the medical hardware gap which Xiaomi has in its hardware cluster.

Health Wearable Platform Scanadu

It is understood that Scanadu has the pedigree of NASA's Ames Research Center. Its medical device Scout, which can measure heart rate, blood pressure, and body temperature, received USD 1.6 million in financing in less than a month and broke the record of lightning-speed financing on the Indiegogo crowdfunding platform.

It has another tool, Scanaflo, which is used to detect the presence of blood cells and other substances in urine, as well as signs of pregnancy and STD infection. It subsequently uploads the test results to the user's smartphone.

On 29 April 2015, the US-based startup Scanadu, which specializes in health equipment, announced that it had secured USD 35 million in its series B financing. This round of investment was led by Fosun International and Tencent and followed by CBC and Singapore's iGlobe Partners. At present, the company's total financing has reached USD 49.7 million. It is reported that this investment will be used to promote and test devices, and will also help Scanadu apply for the FDA's commercial handheld medical device license.

Ybrain for Patients with Alzheimer's Disease

Korean company Ybrain is a start-up company that specializes in neuroscience, which was established in 2013. It especially developed a headband for patients with Alzheimer's disease, which emits electronic signals through two sensors embedded at the front to stimulate their brain activity. At the same time, current tests show that the therapeutic effect of the device is 20% to 30% higher than existing oral medication.

In 2014, Ybrain received USD 3.5 million in its series A financing, and a total financing amount of USD 4.2 million. Its co-founder Seungyeon Kim said the funds would be used for clinical trials and the production of wearable devices.

Glooko Diabetes Management Platform

Glooko, which was founded at the end of 2011, allows users to send data from the blood glucose monitor through the connection between mobile phones and an app for recording, tracking, and statistical analysis purposes. In 2014, Glooko received USD 7 million from Samsung in its series A investment.

In 2015, Glooko again received a new series of financing worth USD 16.5 million. Investors include the well-known American hardware manufacturer Medtronic, whose insulin pumps take up nearly 60% of the market share in the United States. Medtronic is also known as the "crocodile" in the diabetes field.

Raiing Medical

Beijing's Raiing Medical is a developer of wireless medical products and it specializes in wireless monitoring solutions of medical physiological parameters based on the mobile internet. Its founder, Zheng Shibin, has worked for companies such as Hitachi in Japan, Tyco Healthcare in the United States, and Weinmann in Germany, and has many years of experience in OEM and ODM.

Its main products include the iThermonitor, a smart thermometer for children, which can continuously monitor children's body temperatures for 24 hours.

In 2014, Raiing Medical received an angel investment of 10 million *Yuan*, led by Beiruan Angel, and followed by Innoangel Fund, Gobi Partners China, PreAngel and China Young Angel Investor Leader Association.

MC10 Portable Data Monitoring Device

MC10 was established in 2008. It is an electronic product research and development company specializing in health monitoring. Its main products include various kinds of lightweight, flexible, and portable medical data monitoring equipment. These electronic products can be implanted in or used outside the human body. Besides this, they can collect and monitor human blood pressure, brain activity, muscles, water content and other information.

In 2012, MC10 received USD 1 million in financing from medical device manufacturer Medtronic, and venture capital firms North Bridge Venture

Partners and Braemar Energy Ventures to develop invisible health monitoring electronic products.

In 2014, the company received USD 20 million in its series D investment, with a joint investment made by Medtronic and North Bridge Venture Partners.

Augmedix

Augmedix is a technology startup company that promotes applications of Google Glass to doctors. Its products and services can directly push electronic health records to the doctor's Google Glass system, and the doctor can then use voice commands within the system to directly obtain important patient information. Augmedix developed an app on the Google Glass platform, which can display patients' basic data, such as their heart rate, blood pressure, and pulse rate on the Glass.

In 2014, Augmedix obtained USD 7.3 million worth of venture capital investments from DCM, Great Oaks Venture Capital, Rock Health, as well as some individual angel investors.

37mhealth: Chronic Disease Management Platform

The 37mhealth is a health service platform that focuses on the management of chronic diseases. Starting with hypertension, it focuses on building a health management service platform that meets the needs of patients with hypertension. Its product is a blood pressure manager app.

After receiving a seed investment of RMB 200,000 in the business incubator Virtue Inno Valley, 37mhealth received RMB 10 million in its angel round of financing in March 2014. The investors include Broadhi Capital, N5Capital, Innoangel Fund, Beiruan Angel, and angel investors.

3.3   Layout of Plans by International Giants

Who is willing to lose the initiative in the face of such a new market potential? For tech giants, they have no other choice but to seize the opportunities and gain a foothold in the market. Wearable health is igniting the initial enthusiasm for industry innovation and discovery in the medical field. Tech giants in China and overseas are putting in much effort in this area. Particularly on the international stage, industry giants such as Apple, Google, and Sony have been eyeing this new field and are ready to make their moves.

From the tracking of fitness indicators to the monitoring of disease data, and from the management of personalized health details to the execution of a plan for systematic medical insurance services, scientific and technological health management is being realized through wearable devices.

Although wearable devices have not made progress beyond the field of front-end information consumer products such as smartwatches and wristbands, and the pace of technology development in conjunction with smartphone applications is still in its infancy, analysts believe that wearable technology has huge potential in subverting the health care model. This is due to an assurance of people's needs for health care. Plus, wearable technology will help patients save medical costs and shorten the diagnosis and treatment process.

However, with the efforts made by international IT giants, ecological factors such as platforms and data analysis in the wearable industry are constantly improving. Below, we briefly introduce and analyze the layout of the plans which the following industry giants have for the wearable medical industry:

3.3.1   Google: Igniting Wearable Health Care

In April 2012, Google ignited the entire smart wearable industry with smart glasses that extend reality. To date, these glasses have yet to find their way into the consumer market. It may appear that Google did not find the product mature enough and is not in a hurry to launch it. It could also be due to a boycott by users due to an infringement of privacy. Nonetheless, the fundamental reason for this is that Google is relying on Google Glass to get access to big background data. This has always been the biggest ambition, and the biggest advantage, of this search engine company.

Winning by building a platform

In addition to Google Glass, which was the hardware product that was the talk of the town, Google also specifically built a platform for wearable devices, called Android Wear.

In 2013, Google acquired a smartwatch company called WIMM Labs, located in California in the United States. This company first developed an Android-based developer platform for wearable display products many years ago. In 2011, it developed a smartwatch centered on this platform. Developers can develop applications that are compatible with this smartwatch through this platform, too.

Google acquired this company not only because it could provide the talent and technology to develop smartwatches, but also because Google valued the Android application platform it developed for the customization of consumers' smartwatch products.

In 2014, Google launched the first truly wearable system platform, Android Wear. From the purpose of this acquisition, it is evident that Google did not plan to win with wearable hardware from the beginning, but wanted to seize the market for wearable devices through big data, by building a platform.

At the Google I/O Developers Conference in June 2014, Google released its health tracking application development platform Google Fit, and officially launched the Google Fit application in October that same year. Google Fit is similar to Apple's HealthKit solution, and it also provides a management center for checking users' health goals and exercise data. Also, third-party health monitors and applications can access this Google platform and input data. The upgraded Google Fit platform provides 100 new fitness activities for users to choose from. In the original version, there were only three types – running, walking, and cycling. The exercises provided in the new version included badminton, basketball, soccer, aerobics, curling, fencing, hiking, skipping, ice skating, kickboxing, water polo, rowing, running on sand, surfing, volleyball, and yoga. The selection of activities can be said to be the most abundant among similar fitness platforms.

Igniting wearable health care

After igniting the wearable industry with Google Glass, Google continued to leverage on this magical pair of glasses to ignite the core of this industry, which was medical health care.

After being "rejected" by the general consumer market, Google Glass immediately returned to serving enterprises. It has a particularly high usage in hospitals.

In June 2014, Drchrono, an electronic health record company based in Mountain View, California, developed a new application for Google Glass and called it the first "wearable health record". Doctors only need to register with Drchrono and they can download and use this app designed for Google Glass for free. They can then use it to record medical treatment or surgical procedures. Certainly, they will need to obtain the patient's consent first. The recorded videos, photos, and notes are stored in the patient's electronic medical records or Box's cloud, which is engaged in collaboration for cloud storage. The patient can refer to this information at any time.

A foreign startup called Brain Power is currently developing a Google Glass app for children with autism. The program can play the most popular cartoon characters or pictures in Google Glass. When children with autism wear Google Glass to communicate with others, Google Glass can display different cartoon images to express the language and emotions on behalf of the children.

If the wearer wishes to make eye contact with the other party, he or she will only need to look at the other party's glasses. Google Glass will then automatically stop playing cartoon images and directly display the other party's face. The wearer can earn points for themselves, just like in the levels of video games that children are already familiar with. Another Google Glass app developed by Brain Power can focus the wearer's attention on the speaker's eyes during dialogue. This was created because research has found that children with autism pay more attention to the mouth of the speaker when communicating with others.

Google Glass can also enhance the sense of reality in surgical operations, provide doctors with full-body image information, and lower the rate of surgical errors. It can scan the QR codes on ward doors, medicine, and medical equipment to update and synchronize the correct patients' medical records. It can also take over most of the paperwork and recording work in the operating room. Google Glass's camera will continuously capture the doctor's actions during an operation, including images and sounds. It will then extract data from the video to fill in information for electronic medical records.

Launch of a smart wristband for cancer treatment

In 2015, Google applied for a patent from the World Intellectual Property Organization for a wristband, which can destroy cancerous cells in the blood. The device, described in the patent as "Nanoparticle Phoresis," can use "external energy" such as ultrasound or radiofrequency to identify cells in the human body that are harmful to health, and automatically modify or destroy one or more of these targets in the blood. These targets include enzymes, hormones, proteins, cells, or other particles. When they are present in the blood, they can affect a person's medical condition or health.

Health care is the core value of smart wearables. Google's latest patent direction is in demonstrating its in-depth advancement in the field of wearable health care. There is no such thing as pain points or firm demand for smart wearables in this field. As long as we have a desire to live well, live more comfortably, or for a longer time, we will draw more people, enterprises, and funds to the field of wearable devices. After all, it is the most suitable key to connect people and smart technology.

What exactly does Google's patent this time round show? If Google Glass was produced to bring smart wearables out of science fiction films and into the public eye, then the purpose of Google's smart wristband is to let everyone truly understand the meaning of smart wearables. The transformation of medical treatment as shown in the patent is sufficient to influence this era.

The most important point to note is that Google's smart wristband is telling us not to bother about issues between mobile phones and smart wearables anymore. I have said early on that it is only a matter of time before smart wearables replace mobile phones, and the focus of the future is definitely on smart wearables, too. If the focus of the mobile internet era is mobile phones, then the focus of the Internet of Things era must be smart wearables.

Moreover, I believe that the wearable health care industry will be stimulated. Whether it is the media or capital funding, there will be a shift of focus to the wearable health care industry, including wearable devices that can be implanted into the human body, such as nanobot technology, which is used to combat and treat cancer.

Charles Zhang had said, "All diseases will be cured in the next 30 years, and our generation will probably live forever." Some people may not believe it, but with Google's patent on its smart wristbands, humans may truly be just a step away from immortality.

3.3.2   Apple: Building a Health and Medical Platform

Hiring experts in medical sensors to develop wearable devices
Since 2013, there have been media reports saying that Apple had begun recruiting experts in medical sensors, including scientists, engineers, biomedical technologists, and glucose sensors and mass fitness experts.

In early 2014, another two accomplished experts in their fields joined the Apple Watch hardware research and development (R&D) team. They are Nancy Dougherty from the startup Sano Intelligence, and Ravi Narasimhan, the Vice President of R&D of the medical device firm Vital Connect.

Nancy Dougherty was originally responsible for the development of hardware products in Sano Intelligence. The flagship product of Sano Intelligence is said to be a miniature blood analyzer that does not require the drawing of blood. Apple hoped that Nancy Dougherty could apply her rich experience in product design and development on their own wearable devices such as the Apple Watch.

Building health management platforms, HealthKit&Researchkit
At the annual Worldwide Developers Conference held on 2 June 2014, Apple released a new mobile application platform HealthKit, which can integrate data such as blood pressure and body weight collected by other health apps on the iPhone, iPad or Apple Watch, and analyze it and provide feedback.

63

In Apple's iOS 8 system, the built-in HealthKit health platform is supported by many third-party applications and fitness tracking wearable devices, such as Jawbone UP, MyFitnessPal, and Withings Health Mate. At present, more than 900 applications can support data integration.

In addition, Apple has reached agreements with several top hospitals in the United States, such as Duke University Medical Center (Duke Medicine), Ochsner Health System, and Mayo Clinic. Recently, the Los Angeles-based Cedars-Sinai Medical Center updated the hospital's electronic medical record system, importing more than 80,000 patients' data into Apple's HealthKit system. This operation is the largest scale of its kind thus far.

What HealthKit does is that it integrates health-related data from various applications and combines it with electronic medical records, to serve as a form of reference for doctors during diagnosis. It also helps to provide auxiliary functions through real-time tracking, such as calling the police. Evidently, this data will be effective for doctors.

On 10 March 2015, Apple's COO Jeff Williams announced the launch of another new medical and health application Researchkit at its 'Spring Forward' event in San Francisco in the United States. The biggest feature of this platform is its open-source framework. Researchkit can collect medical data of users and conduct effective research by collaborating with experts and medical institutions.

Jeff Williams explained the original intention of launching ResearchKit at the press conference. Its main purpose is for medical research. This is because, at present, there are still many limitations in medical research, such as the difficulty in recruiting volunteers to participate in research and the problem of the requirement of high precision in data collection for medical research. Still, data collected from survey questionnaires is too subjective.

Apple's idea is to tap into the 700 million iPhones currently in use worldwide to record user health information and then use it for relevant medical research.

In terms of privacy protection which everyone is particularly concerned about, Jeff Williams also clarified immediately that users will be informed of the risks of participating in the research activity and asked if they are willing to share personal data with other researchers and partners when joining a

ResearchKit research activity. Hence, the final decision lies entirely in the hands of users.

At present, Researchkit has developed the first batch of its five apps in collaboration with several authoritative institutions around the world. They are the Parkinson mPower Study app for Parkinson's disease, GlucoSuccess for diabetes, MyHeart Counts for cardiovascular disease, the Asthma Health App for asthma, and Share the Journey for breast cancer.

These apps will quantitatively track and analyze users' physical conditions. For example, GlucoSuccess, a collaboration with Massachusetts General Hospital, will remind users to complete the following five steps daily:

1) Measure weight
2) Track exercise (which means that one will need to bring along the iPhone and open the app during exercise daily)
3) Answer two questions daily, including sleep quality and self-monitoring of foot health
4) Monitor food intake (the free app "Lose It!" is meant to be used together)
5) Collect data of Hemoglobin A1C

Recently, according to foreign media reports, Apple intends to use the ResearchKit app to collect DNA test results. The data will mainly be collected by research partners and stored in the cloud online and will be open for use in medical research. For example, the University of California has been collecting DNA from expectant mothers to study and determine the causes of premature birth.

Comparing ResearchKit with HealthKit, one will find that there is a big difference between the two in the user group and the positioning. ResearchKit is mainly used for medical research, while HealthKit is more of an individual health consultant. Besides, in terms of the number of apps, HealthKit will always have far more of them than ResearchKit. Although not all the data in HealthKit may be used for medical research, ResearchKit can leverage on HealthKit and attain great success.

The medical industry in the future will also be gradually transformed in these two directions. Medical data can be collected and integrated into health management platforms, and then further researched and analyzed through platforms like ResearchKit to obtain accurate and effective recommendations for prevention and treatment. These are then going to be relayed back to medical experts and users to form an effective closed-loop for the medical treatment process.

3.3.3   Intel: Big Data Diagnosis and Treatment

Diane Bryant, Vice President and General Manager of Intel, said at a developer's forum in 2014, "Our goal is to jointly sequence all genes and discover the genes that cause tumors and a treatment plan that will prevent the growth of tumors through personalized precise treatment. We will achieve this goal by 2020."

Currently, Intel's main medical projects include a collaboration with Oregon Health & Science University's Knight Cancer Institute to build large-scale supercomputer cluster systems to handle tasks such as genetic identification of tumors.

In August 2014, Intel announced a collaboration with the Michael J. Fox Foundation to research Parkinson's disease, in which they would use custom-made wearable devices to track the health of Parkinson's disease patients, and the data collected would be processed by advanced "big data" computers. This chip giant hopes to solve medical problems with big data.

At present, an estimated 5 million people worldwide have been diagnosed with Parkinson's disease. Parkinson's disease has become the second-largest neurodegenerative disease in the world, second only to Alzheimer's disease.

Within the Parkinson's disease research project, Intel has begun clinical trials on 16 Parkinson's disease patients and nine healthy people. In the next few years, Intel plans to expand this number to a thousand.

Specifically speaking, the trial got Parkinson's disease patients to wear a smartwatch, which is made by Pebble. The watch was used to monitor the patients' sleep and movement to detect convulsions caused by Parkinson's disease. The data was then transferred to Intel's software platform, and the researchers were able to identify certain symptoms not recognizable by the naked eye.

In the second phase of the project, the Michael J. Fox Foundation allocated funds to explore patients' responses to medication. Researchers could monitor the patients' conditions through wearable devices.

Using big data to monitor patients' conditions can be said to be a major trend in the future of the medical field. The accuracy of the data will bring about profound changes to medical research. For example, remote operation mode like Intel's trial is suitable for large-scale clinical trials. Even patients who are in remote areas and not in hospitals can be monitored. This greatly saves time and medical costs for both parties.

Secondly, when unified scientific standards are established, the accuracy of the feedback analysis of data will exceed that of the traditional diagnosis by making observations, listening to breathing, asking about symptoms, and taking the pulse at once. With wearable devices, doctors can assess the frequency and severity of symptoms more accurately, then test the efficacy of new treatment methods. "As more and more devices are launched on the market, we can collect objective indicators and judge the efficacy of new treatment methods," said Sohini Chowdhury, Senior Vice President of the Research Partnerships team at Michael J. Fox Foundation.

Ron Kasabian, Intel's General Manager of Big Data Solutions said that Intel's Datacenter and "Connected Systems" business units are exploring the health industry. "We are studying methods to retrieve data from devices in real-time. We can improve our research by mining data, and to better understand patients' behaviors and their health conditions."

For Intel, the decline of traditional PCs has forced it to re-assess its entry point to the market. The arrival of the wearable devices era and its stimulation of the medical industry has presented an opportunity to Intel, in which it can use its advantage in big data to create new medical prospects with wearable devices.

3.3.4   Facebook: Building a Medical Community

Facebook is a social networking site founded in the United States, launched on 4 February 2004. Its main founder is Mark Zuckerberg, an American.

According to Facebook's financial report for the second quarter of 2014, the total number of monthly active users of Facebook Inc. has exceeded 2.2 billion. Among them, 1.3 billion are monthly active users of Facebook, 500 million are WhatsApp's, 200 million are Instagram's and 200 million are Messenger's monthly active users.

The company's Chief Executive Officer Mark Zuckerberg and Chief Operating Officer Sheryl Sandberg also released some data on the company's operations, including:

As many as 12 billion pieces of information are disseminated on the Facebook platform every day.

There are up to 1.3 billion monthly active users of Facebook.

One billion searches are made by users on Facebook every day.

About 1 billion people use Facebook on their mobile phones.

About 829 million people use Facebook every day.

On the Facebook platform, there are about 800 million followers of public figures.

Another 650 million people use Facebook on mobile devices every day.

About 30 million small companies are using Facebook.

There are 1.5 million companies posting paid advertisements on Facebook.

In addition, the data also shows that Americans spend 40 minutes on Facebook every day. That is one-fifth of the time they spend on mobile devices. Social networking sites in the mobile Internet era is slowly disrupting our traditional way of socializing. At present, the amount of communication and information acquisition performed online has far exceeded to that offline. This is also changing the lifestyles of every one of us.

As social networking platforms accumulate such a huge user group, they can begin to tap into the value of this group in all aspects anytime. When the social networking platform like Facebook saw the IT industry leaders flocking to the medical industry, it also began to take action.

Facebook is ready to enter the medical industry. And this began with a small story:

Eric Topol, Director of the Scripps Translational Science Institute (now Scripps Research Translational Institute), wrote a story in his book about digital medical, healthcare, in which a mother posted photos of her ill son on Facebook, and her friends began commenting on the photos. Three people called the mother, including her cousin who was a pediatric cardiologist. These people told her that her son might have Kawasaki disease, which is a rare genetic disease. The mother then called her doctor and told the latter that she was on her way to the hospital because she felt her son was seriously ill.

"What was I going to say? Three of my Facebook friends think my kid has an extremely rare childhood auto-immune disorder which I just read about on Wikipedia, and since they all contacted me after I posted a photo of him on my wall, I'm going? It seemed... wrong!" This mother named Deborah Kogan wrote on Slate, a famous online magazine. Later, the doctor told her: "You know what? I was just thinking it could be Kawasaki disease. Bravo, Facebook."

The social nature of Facebook spreads information wider and faster, thus allowing this mother to inadvertently obtain effective reference information for her son's condition. This can be said to be beyond everyone's expectations, including that doctor who made their remark. This is exactly where the huge potential lies for Facebook in the medical industry. In addition, we can take a look at another example:

In May 2012, Facebook added the option of an "Organ Donor" status to its Timeline. "An average of 18 people per day die waiting for an organ transplant because there simply aren't enough organ donors to meet the need. Medical experts believe that broader awareness about organ donation could go a long way toward solving the crisis of a persistent lack of organs for matching," wrote Facebook on its blog.

Facebook will thus provide the corresponding registration forms for non-organ donors to fill in, and users can choose whether or not to disclose their acts of organ donation.

When Facebook added registration information for organ donation to its timeline, more than 57,000 people declared their intention to become donors, and 13,000 people officially joined in and became donors in a day. According to data provided by researchers at Johns Hopkins University, this is 21 times the registration rate of ordinary organ donations.

According to the American Journal of Transplantation, on the day Facebook launched the organ donation page, a total of 13,054 people registered to become organ donors across the United States. That is 21 times the average daily count of 616 donors. On the first day after Facebook announced the above news, about 6,000 users in 22 states in the United States registered as organ donors. Before this, there was only an average of 360 people who registered every day. In the days that followed, the momentum of this surge in registrations slowed down, because Facebook would not constantly remind users that they have this option. However, experts who researched the relationship between digital tools and health care said that social experiments have shown that Facebook can indeed have an impact on public health.

"Facebook has such a large and powerful platform, which can be deployed in the medical and health care industry," said Eric Topol.

Facebook had already won the praises of doctors through this act. Soon after, it created online "support communities," through which users can post about various diseases on the social network. In addition, a small team at Facebook was also considering creating a "Preventive Hare" application to help people improve their lifestyles.

Facebook wants to leverage its huge community to create a platform for patients, doctors, and other people who are concerned about health care to consult one another, communicate and interact. For instance, through this platform, patients can share their experiences of overcoming diseases and ask about their conditions, while doctors can exchange information such as a proportion of complicated diseases, clinical diagnosis and treatment experiences, and epidemic transmission routes and prevention methods.

In 2014, Facebook held several meetings with medical industry experts and entrepreneurs and it planned to establish a research and development department to test new health applications. Facebook is currently at the stage of gathering information. Not long ago, Facebook CEO Mark Zuckerberg and his wife donated USD 5 million to the Ravenswood Family Health Center in East Palo Alto.

Users can now add information such as overcoming a disease, losing weight, suffering a fracture, or removing a stent to "life events" on Facebook. This is a sub-category under the "Health and wellbeing" category. The continuous exchange of abundant health care information will allow Facebook to fulfill its potential in the social aspect fully and in a more valuable way.

3.3.5   Microsoft: A Combination of Software and Hardware

In June 2014, Microsoft announced its collaboration with the medical technology company Becton Dickinson (BD) to establish a new incubator specifically for health care and medical technology startups at the Microsoft Venture branch in Tel Aviv, Israel. This would be the only Microsoft incubator focused on the medical technology industry.

Thus far, Microsoft Ventures offices around the world have incubated many successful companies, including some startups in the medical and health care industry. At present, over 200 startups have been incubated by Microsoft Ventures round the world, of which a few dozen are medical-related enterprises. The Microsoft Ventures Accelerator based in China has had six runs so far, and over 100 startups have joined the accelerator. Among them, six are related to medical and health care, and they cover many fields including chronic disease management, health management, and maternal and infant health.

Microsoft had invested USD 250 million in WebMD, a health information website as early as May 1999. As such, it had begun to enter the medical and health care industry through mergers and acquisitions.

In 2006, Microsoft stated that as part of a new collaboration agreement with the non-service medical institution MedStar Health, it would acquire the medical database software Azyxxi developed by the latter's Washington Hospital Center. After the acquisition was completed, Azyxxi would be launched in the global medical market. That is also the first time Microsoft entered the field of medical information technology.

In 2007, Microsoft's then CEO Steve Ballmer announced at a conference for medical and health care information in New Orleans, that Microsoft would acquire Medstory, a search engine for health information.

In 2009, Microsoft made another announcement that it would acquire Sentillion, a software manufacturer specializing in the medical and health care industry, and would combine Sentillion's products with Microsoft's Amalga UIS. This would allow professionals in the health care industry to obtain various IT applications and patient information more easily and thus provide patients with better medical services.

Development of medical wearable hardware
In the aspect of smart hardware, Microsoft has also had a significant amount of development and trials.

1. Alice Band
According to foreign media reports, Microsoft was working with the Guide Dogs for the Blind, a guide dog training school, as part of the United Kingdom's Cities Unlocked project, which is one of the seven Future Cities Catapult which the UK government had launched for specific groups in the community. This project was initiated by the UK's Technology Strategy Board, and the UK government had planned to invest 1 billion pounds worth of research and development funds in this specialized field over the next five years to globally launch its leading innovation activities.

For this project, Microsoft was developing wearable devices with special functions for people with disabilities. These were aimed to help them to better live independently.

The wearable device named "Alice Band" developed by Microsoft can help the blind and those with weak eyesight to live like able-bodied people. The working principle of the Alice Band is to obtain information from sensors installed in buildings or train cars through bone conduction 3D headphones installed on the user's head, and convert them into 3D audio prompts for navigation. This is a very considerate design for the blind and people with weak eyesight, who rely heavily on sounds to obtain information.

Alice Band can help the blind find their way on congested stairs and escalators in the busiest interchange stations and airport entrance gates in the UK. The blind can also use this service in banks or shops.

2. Microsoft Band

At the end of October 2014, Microsoft released its smart wristband, the Microsoft Band. This device would be connected to Microsoft's new health fitness tracking service, Microsoft Health, which has been discontinued as of 31 May 2019. The Microsoft Band had 10 built-in smart sensors, which could monitor the user's sleep and heart rate, calories burned, and other data during exercise in real-time. Through collaboration with well-known gyms, Microsoft Band could provide users with reasonable fitness programs to help them achieve their health goals.

Besides, the wristband could allow users to view their health data at a glance through calendar notifications and email previews. It could also take notes and schedule reminders using the Cortana voice assistant.

3. Microsoft Health

Apple has HealthKit, Google has Google Fit and Samsung has SAMI. Watching its peers launch their platforms, the tech giant Microsoft also jumped on the bandwagon and launched its own health management platform, Microsoft Health.

The Microsoft Health cloud service platform, which has been discontinued as of 31 May 2019, was used for consumer and industry storage. It could integrate data collected from different health and fitness devices and store the data securely in the cloud. Users could compare and analyze the data that was stored in the Microsoft Health cloud with the data they had obtained from different devices. The data included step count, calories, and heart rate. Users could then draw valuable conclusions through Microsoft's "Intelligence Engine," such as the type of exercise that burned the most calories, and then consolidate the health and fitness data.

The "Intelligence Engine" in Microsoft Health could suggest reasonable workout plans for users and offer advice for their lifestyles, such as whether having breakfast would make them run faster, or whether the number of meetings attended in the day would affect their quality of sleep.

Microsoft had achieved cooperation intentions on Microsoft Health with equipment and service providers including Jawbone UP, MapMyFitness, MyFitnessPal, and Runkeeper. Microsoft also had plans to provide options for users to share data with health care providers through the connection between Microsoft Health and HealthVault.

4. HealthVault

In 2007, Microsoft launched HealthVault, a platform that can store and manage users' health information. This was shut down on 20 November 2019. On the HealthVault personal health management platform, users could maintain their health records online as long as they applied for a personal health account online. The platform was like a personal information safe with an open interface, which could exchange data with third-party equipment manufacturers and insurance companies. Users could decide what information content to upload and to whom the information access could be given.

In other words, users could upload data measured on other devices onto this platform, and HealthVault would give users a series of relatively accurate and objective solutions after integrating and analyzing the data from all sources.

When other industry leaders were launching their health management platforms, Microsoft was constantly upgrading and perfecting its own platform. This secured a large number of active users for the various Microsoft wearable devices that it later launched. Consumers could then upload biological data indicators such as blood pressure, respiratory value, and blood glucose levels from 233 third-party devices onto HealthVault, and could exchange data with more than 160 third-party applications.

5. Xbox Fitness

In September 2013, Microsoft launched Xbox Fitness, a personal fitness service, while releasing its new generation of gaming console Xbox One. The service was discontinued on 1 July 2017.

Xbox Fitness featured fitness training through gaming. Microsoft's renowned Kinect greatly enhanced the charm of Xbox games. Users could play dancing and sports games by moving their whole bodies, without controllers. This increased their sense of immersion in the game. While enhancing the fun of the game, Xbox Fitness also had an embedded function that could track the quality of the workout by identifying data such as users' heart rate and muscle strength. It could then dynamically adjust the workout plan based on the users' exercise history and past effects.

To ensure the quality and professionalism of the fitness training, Xbox Fitness introduced teaching videos of many famous fitness brands, including P90X® (by Tony Horton), INSANITY® (by Shaun T), Jillian Michaels, and Tracy Anderson. Xbox Fitness was attractive to users because they could use these fitness video teaching materials directly and follow the trainer on the TV. Kinect would capture the user's real-time images and motion data, and display them on the screen in real-time, allowing the user to understand his or her fitness condition. What was more convenient was that fitness trainers did not need to develop new teaching materials specifically for Xbox Fitness. It only required some fine-tuning of the sensors and feedback from Microsoft.

Xbox Fitness was also integrated with Xbox Live, which allowed users to enhance their enthusiasm for exercising by competing with the user's friends. There were more than 1.6 million users of Xbox Fitness.

It is evident from this series of projects developed by Microsoft that Microsoft had spared no effort in the medical and health care industry. We can understand the layout of Microsoft's plan more intuitively from the following figure (Figure 3-9).

For now, Microsoft undoubtedly owns the world's best research and development (R&D) institutions in hardware equipment, but unlike other industry leaders, its market performance pales in comparison with its strong R&D capabilities. However, recently in the field of smart wearable devices, Microsoft has begun to make frequent moves. It will emphasize market expansion in its next steps. For an IT giant like Microsoft, it has no problem stimulating market vitality, as long as it has sufficiently good products.

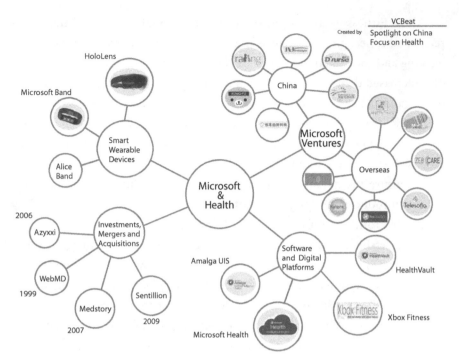

Figure 3-9 Layout of Microsoft's plan in the internet
medical and health care industry

3.3.6   Samsung: Modular Health and Medical Platform

Creating SAMI, an open-source platform for wearables
In 2013, Luc Julia, a former Apple Siri developer, moved to Samsung to serve as
the Vice President of the latter's Innovation Center. He was mainly responsible
for the Samsung Architecture for Multimodal Interactions (SAMI) project.

In May 2014, Samsung held a product launch conference called "Voice of
the Body" in San Francisco, the United States, and launched a new open-source
platform SAMI. All data collected in the future would be saved on the SAMI
cloud platform. In terms of security, Samsung said that all the data on the
cloud would be protected from market abuse. The biggest difference between
Samsung and other IT giants is that Samsung would give all developers access

to the API of this platform, which would allow this platform to collect and develop big data more effectively.

At the product launch conference, Samsung also launched a smart hardware device, the Simband. The concept picture below is one of its prototypes, which has multiple built-in sensors, such as a visible light sensor that can be used to detect the amount of light absorption in the skin (to obtain pulse rate and other data ultimately) or an ECG (electrocardiogram) sensor that is integrated with the wristband.

According to Samsung, all components of the Simband can be replaced, and hardware developers can even insert self-created hardware modules into Samsung's Simband, which makes customization easier. Samsung also claimed that the power consumption of this product is very low, and the size of the device would not be too big (the surface of the prototype we have seen is only as large as an SD card).

Samsung Smart Health Ecosystem
At the Boao Forum for Asia in 2015, Samsung launched various exhibition activities such as exhibition halls, forums about smart health and wearables, and large-scale outdoor advertising. In the Samsung exhibition area, the Samsung smart health ecosystem composed of bioengineering processors, Simband, Gear, and the S Health application was displayed to let participants understand the concept of smart health and its value to customers.

Samsung played videos of daily health management such as the monitoring of exercise, diet and sleep, and demonstration of the measurement of blood pressure, pulse rate, and blood glucose levels on large screen displays. It is worth mentioning that the Galaxy S6 unveiled at the Mobile World Congress would be connected to the smart health products, and visitors could directly try it.

Samsung said that it would accelerate the arrival of the digital health era by collaborating with the IT industry and medical institutions and using IT tools to build its own health management system. In fact, Samsung had begun working with the American diabetes management companies and was looking into entering the mobile phone health care market. In the future, Samsung would also strengthen collaborations with insurance companies, pharmaceutical factories, and hospitals, in search of the best way to reduce medical expenses in aging communities.

Young Sohn, the President of Samsung Electronics said, "Samsung Electronics' goal is to provide consumers with solutions for managing physical and mental health through smartphones. It is our goal to use smartphones, wearable devices, bioengineering processes, and health care software application systems and services of sensors to build a mobile phone health care ecosystem."

3.3.7 Qualcomm

Due to the slowdown in the growth of the smartphone business globally, Qualcomm, the world's largest mobile chip manufacturer, has set its sights on the most popular industries at the moment, including smart health care, smart wearables, and smart homes.

At the Boao Forum for Asia in 2015, Qualcomm's President Derek Aberle said in an interview that they had invested USD 33 billion in the entire Internet of Things industry and established a dedicated investment department.

In terms of mobile healthcare, Derek said that Qualcomm hoped to reduce medical costs through technology, and provide convenient services for patients. To this end, Qualcomm would collaborate with many pharmaceutical manufacturers. They had already established collaborations with some of them, including the pharmacy chain Walgreens and the pharmaceutical company Novartis. Besides, they would use their wireless technology to provide remote monitoring for patients to get the corresponding data.

In terms of smart wearables, Derek said that Qualcomm would extend its previous strategic partnership to provide professional wearable chips for wearable devices' manufacturers. In the smart home industry, Qualcomm would develop a project called AllJoyn, with the main purpose of achieving interconnection between homes.

At the end of 2014, Qualcomm announced that it had funded the establishment of the China Center for mHealth Innovation (CCmHI), a research organization sponsored by The George Institute for Global Health, a non-profit research organization. Through this organization, Qualcomm would realize its plans for a digital medical system and electronic medical records, real-time diagnosis results at medical treatment stations, and community systems of regional health care management agencies being linked together.

In the future, patients would be able to transfer data to cloud servers through software applications for self-treatment.

It is reported that CCmHI would first conduct an analysis of the current status of China's digital health policies, laws, standards, projects, and research activities, and use it as a basis to drive other work. Qualcomm said that CCmHI planned to conduct a test for the development of a mobile health project for chronic disease management, "to implement larger-scale projects in the future."

Qualcomm's biggest advantage lies in its chip manufacturing and wireless transmission technologies, which are at the core of the hardware and software technology for wearable devices. With other smart hardware devices slowing down in their development at the moment, channeling resources to the wearable devices industry will be Qualcomm's next move in its largest strategic plan.

CHAPTER 4

The Driving Force behind the Wearable Health Care Market

4.1 Global Aging

According to the classification criteria established in *The Aging of Populations and its Economic and Social Implications* published by the United Nations in 1956, a population is considered to be aging if more than 10% of the total population is ≥60 years of age, or more than 7% of the total population is ≥65 years of age. The life expectancy of populations has changed dramatically in the 20th century. Compared with 1950, the average life expectancy has been extended by 20 years to 66 years, and it is expected to be extended by another 10 years by 2050. This great progress in population structure and the rapid growth in population in the first half of the 21st century mean that the number of people over 60 years of age will increase from approximately 600 million in 2000 to nearly 2 billion in 2050.

The forecast released by the United Nations in 2005 shows that the proportion of the world's elderly population over 60 years of age will rise from 10.0% in 2000 to 15.1% in 2025, and 21.7% in 2050. At the same time, the proportion of the elderly population over 65 years of age will increase from

6.9% to 10.5%, and 16.1%, accordingly. The median age will thus rise from 26.8 years old to 32.8, and 37.8.

ABI estimates that by 2017, applications for remote patient monitoring and online professional health care will account for 20% of the wearable wireless device market. Some analysts believe that if extended to the entire health care industry, fitness and medical wearable devices will account for 60% of the wearable device market.

IMS, a world-renowned medical consulting agency predicts that by 2050, one in every five people in the world will be over 60 years old. Hypertension, late-onset diabetes, heart disease, and other chronic diseases that require monitoring will become the driving force for the growth in demand for wearable devices.

4.1.1   Current Aging Situations in Other Countries

United States

At present, the elderly aged 65 and above account for about 12.5% of the national population in the United States, and this proportion will reach 20.7% in 2050. Among them, there will be over 18 million people aged 85 and above, which is nearly six times the number in 1995.

In terms of life expectancy, according to the US Census Bureau, the life expectancy was 47 years for the population born in 1900, 68 years for those born in 1950, and 77 years for those born in 2000. The current life expectancy of 65-year-old men and women is 16 and 19 years respectively, while it was 11 and 12 years just in the early 20th century. The life expectancy of 65-year-old men and women has increased by about 40% and 60%, respectively.

In terms of medical needs, about 40% of the elderly in the U.S. need to spend some time in hospitals or other care institutions, and about 4.5% of the elderly will spend the rest of their lives there. Moreover, according to the current growth rate estimates, the number of Americans living in nursing homes will amount to 3 million in 2030, about double that in 1995. All these indicate that the aging of the population will bring unprecedented pressure to the country in all aspects, especially in medical construction.

Japan

According to data from the Ministry of Health, Labor and Welfare of Japan, Japan's population fell by a record high of 268,000 in 2014. In 2014, 1 million people were born in Japan, which was a decrease of 2.8% compared with the previous year, while the death toll rose to 1.27 million.

The current status of the Japanese population can be summarized as "super-aged." The country's population of elderly people above the age of 65 has reached a record high of 32.96 million. According to relevant data, one in eight people in Japan is above the age of 75. Japan is a country of longevity. In 2003, the average life expectancy was 78.36 years for men and 85.33 years for women, and 19.24% of the population was over the age of 65. These numbers are still on the rise currently. According to estimates by the National Institute of Population and Social Security Research in Japan, the proportion of the population that is above the age of 65 will exceed 30% by 2024, and amount to 33.4% by 2035.

As the country with the longest life expectancy in the world, Japan began to have negative population growth in 2010. Its population will reduce to 67.366 million by 2100, which is only slightly more than half of that in 1998. Japan's current fertility rate is around 1.4. By 2060, Japan's population will reduce by one-third from the current count of 127 million, and two-thirds by 2110. Even if the fertility rate rises rapidly to the equilibrium point of 2.07 by 2030, the population of Japan will continue to decrease in the next 50 years to less than 100 million.

This has a direct impact on the entire country. On one hand, the labor force will be greatly reduced, on the other hand, the social welfare expenditure will soar, which will further increase the Japanese government's debt.

Australia

According to the Australian Bureau of Statistics, the proportion of the country's population that is aged 65 and above increased from 11.1% on 30 June 1990 to 13.6% on 30 June 2010. Over the same period, the proportion of the total population that is aged 85 and above has doubled, from 0.9% in June 1990 to 1.8% in 2010.

South Korea

According to Statistics Korea, South Korea had a population of 48 million in 2006. It is estimated that between 2045 and 2050, South Korea's population will decline at an average annual rate of 1%. By that time, South Korea will become the country with the fastest population reduction in the world. By 2050, the number of people above the age of 65 will account for 38.2% of South Korea's total population, when the world's average is only 16.2%. This will make South Korea the world's "oldest" country. Analysts in South Korea warned that low birth rates will lead to a reduction in the domestic labor force, which will aggravate the aging of its society and put a heavy burden on the government to provide for the elderly.

European Union

The EU Green Paper on population published by the European Commission shows that the total population of the EU will amount to 468.7 million by 2030, and the labor shortage will be 20.8 million. In 2030, two people in the working-age population (those aged 15–64) will have to feed one person in the non-working-age population (those aged 65 and above). Between 2005 and 2030, the EU's elderly population above the age of 65 will increase by 52.3%, while the proportion of its population aged 15–64 will drop by 6.8%. To make up for the employment gap caused by aging, the average employment rate in the EU must reach 70%. However, the average employment rate in the EU between 2001 and 2003 was only 63%.

4.1.2   In China, the Population Ages before Becoming Wealthy

According to the most recent data released by the Vice Minister of Civil Affairs, Dou Yupei, at the press conference for the State Council Information Office, there are more than 200 million elderly above the age of 60 in China today. This accounts for 14.9% of the total population. This proportion is significantly higher than the United Nations' standard of 10% for a traditional aging society. Last year, the China National Committee on Aging predicted that China will reach the peak of aging in the next 20 years (Figure 4-1).

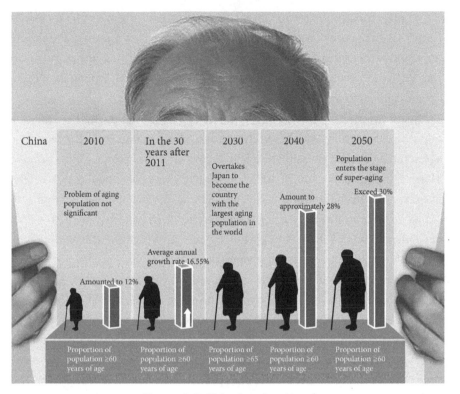

Figure 4-1 China's aging trend

As the family planning generation has begun to age, there are more and more elderly people who either have no children or have lost their only children. In 2012, there were at least 1 million families who lost their only children in China, and this number continued to increase by about 76,000 each year.

According to the *Report on the Development of China's Aging Society and Pension Security* (2014) jointly released by the Research Center of Employment & Social Security of Tsinghua University and *China Economic Weekly*, in 2013, the "Medical Security Development Index" rose to 63.5 from 62.7 points in the previous year and it was the only index that passed. The "Pension Development Index" improved slightly, but it still did not reach the passing mark. It is particularly noteworthy that the "Aging Society and Grey Hair Economic Development Index" failed, with a score that was lower than the previous year's.

In 2013, the proportion of China's population over the age of 65 increased from 9.4% in the previous year to 9.7%, and the dependency ratio of the elderly population (the ratio of the elderly population to the working-age population) was 13.1%. However, if calculated according to the actual dependency ratio, China's aging situation is even more severe.

In addition, if working-age students, the unemployed, low-income population and those who have retired early are taken out from the working-age population, the aging situation will seem even more serious. China became a deep aging society before 2010 (the dependency ratio was about 5:1), whereby five taxpayers from the working-age population supported one elderly person. China may become a super-aging society earlier than expected in 2020 (with a dependency ratio of about 2.5:1), whereby 2.5 taxpayers from the working-age population will support one elderly person.

Moreover, many elderly people are plagued with chronic diseases. According to statistics, the prevalence of chronic diseases is 169.8‰ in China's total population and 540.3‰ among the elderly. Among the population above the age of 65, most people suffer from cataracts, depression, diabetes, and Alzheimer's disease, and 42% of the elderly suffer from two or more diseases at the same time. Among them, the prevalence of hypertension, heart, cerebrovascular, and respiratory diseases are higher, and there is an increasing trend year on year. It is evident from these statistics that chronic diseases have become the biggest "murderer" of the elderly's health.

In addition, chronic diseases are becoming a public health problem.

4.2   The Prevalence of Chronic Diseases in the General Population and a Younger Age Group

Chronic disease is a general term used for diseases that are not infectious but accumulate over a long time to form disease morphological damage. Examples include cardiovascular diseases, diabetes, malignant tumors, and chronic respiratory diseases. Chronic diseases often cause huge harm. If prevention and treatment get out of hand, chronic diseases can cause economic harm and endanger lives.

On 19 January 2015, a new report from the World Health Organization showed that non-communicable chronic diseases such as cancer, cardiopulmonary diseases, stroke, and diabetes are still the leading causes of death in the world. Many deaths occurred prematurely and could have been avoided.

WHO data: In 2012, there were 38 million deaths from non-communicable chronic diseases worldwide, of which 8.6 million occurred in China. Every year, 3 million people in China die prematurely, that is before the age of 70, from certain preventable diseases.

About 39% of men and 31.9% of women died of chronic diseases in China died prematurely. The World Bank estimates that between 2010 and 2040, China can generate an economic benefit of USD 10.7 trillion by merely reducing the death rate from cardiovascular and cerebrovascular diseases by 1%. In China, more than half of the men smoke (though only 2% of females are smokers), more than four-fifths (83.8%) of teenagers aged 11–17 are not physically active, and nearly one-fifth (20.2%) of adults have hypertension.

Let us first understand the incidence of chronic diseases among Chinese residents in the 10 years between 2003 and 2013 from the statistical charts in the *2014 Statistical Yearbook for Health and Family Planning in China*.

The average incidence of diabetes in China increased by nearly seven times in 10 years, of which the incidence among the urban population increased by nearly three times, while that of the rural population increased by as much as 10 times. One can predict that the rural areas will be the hardest hit by chronic diseases in the future. This is closely related to the changes in the diet and lifestyle of the rural population in the past 10 years.

The average incidence of hypertension in China increased by about six times in 10 years, of which the incidence among the urban population increased three times, while that of the rural population increased by about eight times.

Currently, there are 260 million people in China who suffer from chronic diseases. In 1998, the number of people who suffered from chronic diseases

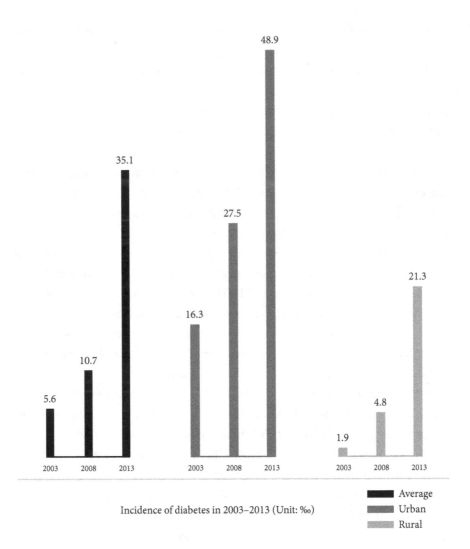

Incidence of diabetes in 2003–2013 (Unit: ‰)

Average
Urban
Rural

Figure 4-2

accounted for only 12.8% of the population, but this proportion has risen to 15.7% in 2008. Chronic diseases have accounted for 85% of the deaths in China, and 70% of the entire medical expenditure.

There are now 98 million people who suffer from diabetes in China, and many reports are showing that China's chronic disease patients are gradually

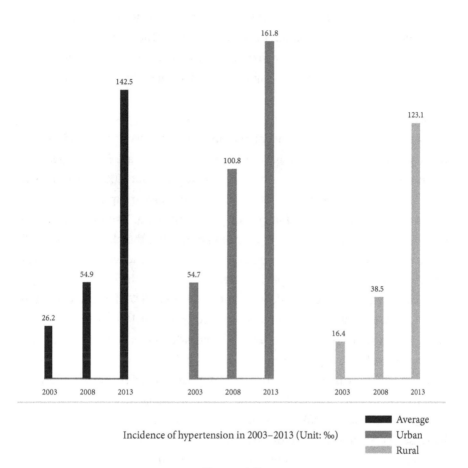

Incidence of hypertension in 2003–2013 (Unit: ‰)

Average
Urban
Rural

Figure 4-3

getting younger. The number of patients over the age of 40 who suffer from chronic diseases has tripled or quadrupled within 20 years.

A comparison of common causes of death in various countries shows that the mortality rate due to cerebrovascular diseases in China is four to five times higher than that in countries like the United States, United Kingdom, and France. This is 3.5 times higher than that in Japan and higher than India, which is also a developing country. The mortality rate due to malignant tumors in China is close to that of the United States, the United Kingdom, and France, but higher than in Japan and India. The mortality rate due to respiratory diseases in China is higher than that in all other countries. The mortality rate due to heart diseases in China is lower than that of India but close to the levels of the United States and the United Kingdom, and significantly higher than that in Japan and France.

In addition, China's chronic diseases are showing a trend of getting "younger." Research shows that more than 65% of the workforce suffers from chronic diseases. The age groups of this group of people are 16 to 60 years old for men and 16 to 55 years old for women. There are about 69% of hypertension cases and 65% of diabetes cases occur in the above age groups.

At the end of 2014, *Wired* magazine published an analysis of the prospects of the wearables market. The article pointed out that the elderly, chronically ill patients, and poor people are actually the biggest beneficiaries of wearable devices, but this group of people has been neglected by the market.

According to a study by Pew Research Center, 45% of American adults suffer from at least one chronic disease, and less than 19% of users with no chronic conditions track their health indicators. This means 40% of American adults who suffer from one chronic disease track their health indicators. Among American adults who suffer from two chronic diseases, 62% of them do so.

The article also pointed out that the sales of wearable medical devices in the United States reached USD 2.8 billion. In the next five years, this number will increase to USD 8.3 billion. If one had taken note of the sales volumes of fitness wristbands and smartwatches this year, it is apparent that even multiplying the number by six will give an amount that is much lower than the USD 6.3 billion worth of sales in the U.S. blood glucose test strip market.

Another phenomenon is that more than half of the fitness tracking device users in the U.S. no longer use the various wearable fitness devices. About 1/3 of these users kept the devices in their drawers or gave them to friends after using them for less than six months.

Obviously, this is a problem that should raise alarm bells in the market, but it also reveals the ultimate value of wearable devices, that is, it can bring solutions to true health care. In face of the current trends of aging populations and the high incidence of chronic diseases worldwide, the wearable device market should probably refocus. We understand that these groups of people are using health management wearable devices like other ordinary people, but the biggest difference lies in that the former will not abandon the health devices after the initial excitement wanes. This is because tracking and monitoring their health indicators can keep them away from the hospital.

For instance, Dr. James Amor, a researcher from the University of Warwick believes that if the elderly can wear smartwatches or smart clothing that can measure heart rates, temperatures, workouts, and other physiological characteristics, the entire monitoring activity will allow the elderly's family members and caretakers to understand their health conditions and daily activities. This can give users greater confidence and reduce their hospitalization rates.

Therefore, this group of people will become the most stable user group in the wearable medical industry in the future. Conversely, they are also the people who truly need wearable medical devices. At the same time, the medical expenses of an aging society plagued with chronic diseases are dragging down the economies of some countries. Whether wearable health care can be approached from a new perspective will affect the development of the entire country's economy.

4.3 The Arrival of the Big Health Era

Feng Lun, Chairman of Vantone Holdings, understands big health as the health management which begins from the time one is still just a gene in the mother's womb. For instance, most diseases such as Down syndrome can be detected within the first three months of pregnancy. Big health management refers to the management of everything from the point of fertilization to the time one enters the Babaoshan Cemetery.

As living conditions improve, medical technology advances, and health awareness is strengthened, the concept of "big health" has become increasingly popular in recent years. Big health is essentially a broad concept of health. In other words, it is the strengthening of public awareness of health that has led to such a concept. This concept mainly revolves around people's clothing, diet, housing, transportation, life, and death. It focuses on various health risk factors and misconceptions and promotes self-management of health. On top of a scientifically healthy life, one should also have the correct health concepts.

In a narrow sense, the big health industry includes medical equipment, medical and health care, and pharmaceutical industries that have the goals of treating diseases and maintaining lives. It also includes industries with the goals of delaying aging, preventing diseases, and maintaining healthy lives, such as health products, functional foods, safe water, and healthy drinks, as well as industries related to the development and healthy environments, such as environmental protection and natural resource industries (Table 4-1).

Table 4-1: Classification of the Big Health Industry

Classification of the Big Health Industry	Scope
Product-oriented	All products are based on health-related concepts such as health food, health tea, health wine, health products, cosmeceuticals, medicinal wine, and health products for daily use.
Service-oriented	Mainly includes the medical industry with medical service institutions as its main body, the health management service industry with personalized health examination and evaluation, consulting services, conditioning, rehabilitation, and safeguarding at its core, health insurance, health risk management industry, and sports health services such as health centers.

Naturally, the concept of big health has become highly regarded, because the health conditions of all citizens have threatened the development and the future of the country.

Lang Xianping, an economist, pointed out that the key factor that would affect China's GDP growth in the future would no longer be a labor productivity issue which is part of the demographic dividend, but a sub-health issue instead. It is common knowledge that a family's income will be reduced significantly if the breadwinners fall ill. Correspondingly, for a country, the more patients there are, and the higher the proportion of adults who are sub-healthy, the greater the impact there will be on GDP growth.

The research found that there are currently 160 million adults with excessive blood lipid levels, 200 million adults who are overweight, and 300 million adults who are smokers in China. The proportion of cancer outbreaks has doubled in the past 40 years. Besides, in the next five to 10 years, 1.4 trillion *Yuan* may need to be invested to build 50 million hospital beds. People's health problems have become one of the biggest obstacles to China's economic development.

The big health era will be closely related to everyone, even the economic lifeline of the entire country. Health comes before the economy, and being healthy will allow a young country to quickly take off.

In October 2013, the State Council issued the *Opinions of the State Council on Promoting the Development of the Health Care Service Industry*, which had highly advantageous policies for the pharmaceutical industry. The *12th Five-Year Plan* proposed a shift in focus in China's medical and health industry development from treatment-based to prevention-based, and from infectious disease prevention to chronic disease prevention. Various policies have contributed to the rapid development of the big health industry.

In November 2011, the State Council issued the *12th Five-Year Plan for Medical Science* and Technology, which proposed the development goal of "cultivating big health industries and developing new types of health products." This is the country's first plan made for the big health industry, and also the first plan made for the industry as a whole.

Whether it is the reality or a prediction of the future based on the situation today, there are telltale signs that the era of big health has come. Everyone would start to care about their health and seek the best health interpretation scheme. But there is a problem, whereby the health aspect of the big health era cannot be fulfilled in the hospital but is integrated into every aspect of people's lifestyles, anytime, anywhere. In other words, it means to cultivate a healthy living concept and habits. This is exactly the strength of wearable devices.

4.4 Favorable National Policies

On 28 September 2013, the *Opinions of the State Council on Promoting the Development of the Health Care Service Industry* (hereinafter referred to as Opinions) was published, which mentioned the term "non-prohibited access." This means that all the areas that are not explicitly forbidden by law and regulations must be open to social capital, and all the areas that are open to local capital must be open to foreign capital. This raised hopes that the "glass door" blocking private capital from entering the medical and health care industry could be broken.

In the Opinions, the State Council also stated that by 2020, it would basically establish a health service system that covers the entire life cycle, is rich in content and has a reasonable structure, and fundamentally meets the health service needs of the masses. The scale of the entire health service industry would amount to over 800 trillion *Yuan* and it would become an important driving force for sustained economic and social development.

In May 2014, the China Food and Drug Administration issued the *Measures for the Supervision and Management of Food and Drug Sales on the Internet* (Draft for Comment), which was planned to be implemented in 2015. It included a new policy that would be implemented to allow the sale of prescription drugs online. Also in May, the State Council issued the *Notice on Issuing the 2014 Major Task List on Deepening the Medical and Health Care System Reform,* which placed the promotion of public hospital reform as its top priority, and set hard deadlines by specifying the time for completion of each task. According to media reports, there would be big moves implemented as part of public hospital reforms in 2015.

On 12 January 2015, the National Health and Family Planning Commission issued *Opinions on Promoting and Regulating Multi-point Practice of Physicians,* proposing to promote a rational flow of physicians, to relax conditions, simplify registration approval procedures, explore the implementation of record management or regional registration, and optimize the policies for physicians' multi-point practices. This would allow a further liberalization of multi-point practices, which is a key part of the medical reform.

On 6 March 2015, the State Council issued the *Outline for the Planning of the National Medical and Health Service System (2015–2020)*, which stated that China would launch plans for a healthy China cloud service and actively apply new technologies such as mobile Internet, Internet of Things, cloud computing and wearable devices. This was to promote health care informatization and smart medical services that benefit all, promote the application of big data in health care, gradually change service models, and improve service and management capabilities.

As remote monitoring of chronic diseases can reduce the overall medical cost, it has obvious economic advantages (for example, a study on diabetic patients in the U.S. shows that remote monitoring can reduce medical costs by more than 40%), which will maximize the use of existing medical resources (waste in medical resources account for more than 30% of medical expenses in China) and address the issue of a serious shortage of medical resources in remote areas. The Chinese government has thus shown clear support for mobile health care.

The relevant national ministries and commissions have issued a series of documents and policies to encourage and support the development of mobile health care. The Ministry of Industry and Information Technology and the Ministry of Science and Technology have listed personal medical monitoring and remote diagnosis as the development priorities in the *12th Five-Year Plan* respectively and provided corresponding financial support. The Ministry of Health has also initiated and sponsored mobile health care demonstration case studies in some cooperative medical service demonstration projects, including mobile medical solutions such as medical record-keeping, disease data, and health quality monitoring.

4.5 Lazy Economy as a Catalyst

Don't want to clean, but love to be clean? Lying in bed but need to get something? Don't want to cook, but don't want to head out for dinner as well... Modern lifestyles have created many lazy people. The demands of these people are always unreasonable, but step by step, they are beginning to be fulfilled. Society progresses to make everything easier and more convenient. In other words, it is grooming more "lazy people."

If one truly wishes to earn some gold from this group of people, one must have an accurate grasp of their consumer psychology and enhance the practical function of the service to satisfy the desires of "lazy people" to be "completely lazy." As a result, a new industry has emerged – the lazy economy.

Jack Ma used many examples in one of his speeches to prove that we have this wonderful world thanks to lazy people. Thomas Edison told future generations that genius is 1% inspiration and 99% perspiration. This is because he was too lazy to think about the real reason behind his success.

Bill Gates, the world's richest man, dropped out of school because he was too lazy to study. Later in life, he was too lazy to remember the complicated DOS command, so he developed the graphical user interface program.

The boss of Coca-Cola was even lazier. Although there was a Chinese tea culture with a long history and the fragrant and delicious Brazilian coffee, he was too lazy. He added some cold water to the artificial sweetener, bottled the product, and sold it. It did not take long before everyone in the world began to drink this cheap liquid, and this made Coca-Cola the most valuable brand in the world.

McDonald's, the world's greatest food and beverage enterprise, also had a surprisingly lazy boss. He was too lazy to learn about the delicacy of French cuisine and too lazy to master the complicated techniques of Chinese food. He just placed a beef patty in between two pathetic pieces of bread and sold it. Owing to this, the M logo can now be seen all over the world. The owner of Pizza Hut was too lazy to put the stuffing in the pie, so he scattered them on the dough and sold it like this. Everyone calls it pizza, and it is more expensive than 10 pies altogether.

Someone was too lazy to climb the stairs, and so the elevator was invented.

Someone was too lazy to walk, and so cars, trains and airplanes were invented.

Nowadays, people's lives revolve around these things invented by these lazy people. As long as these products can make people lazier, there will be a huge market for these things. This is where the opportunity lies for wearable devices. Firstly, it makes the purchase of items easier and safer than now. In the future, people will no longer need to remember those annoying passwords, link their identities, and set up verification, yet still worry about the security of their accounts. Wearable devices have the function of identifying unique characteristics of the human body quickly and effectively, which is safer than even the most complicated password or verification process.

Secondly, there is no need to make an appointment in advance or join long queues to register to see a doctor. Patients can communicate with doctors anytime and anywhere through wearable devices, and doctors can use the various health data sent back by the wearable devices previously to diagnose the patients and offer guidance for a healthy lifestyle.

Wearable devices will become the personal assistants, doctors, and fitness coaches in your life. They can even read your subconscious and tailor your living environment to your needs before you even realize it.

CHAPTER 5

The Wearable Health Care Ecosystem

5.1 Big Data Medical Platform

Hu Xiaoming, President of Alibaba Cloud, believes that cloud computing and big data are two sides of the same coin. The hospital of the future is likely to be a hospital of data, and the hospital management of the future will also focus on data management. This includes data of patients and their family members, and it will manage the entire life cycle of individuals. This data can allow the entire medical service to overcome limitations due to geographical locations, diagnosis, and treatment time. The data also makes it possible to achieve breakthroughs for certain complex diseases.

We are striving towards such a future, in which massive data processing will cover all aspects of our lives. Particularly with the advent of wearable devices and the mobile internet era, the medical field will usher in profound changes that create more possibilities for doctors, patients, and medical research. At the same time, we are also going through a thorough transformation, in which we transform from understanding the world based on a small amount of information to focusing on new horizons based on the continuous collection of unlimited data.

5.1.1 Big Data

In ancient Rome, humans could not make data calculations above 10,000. When Apollo landed on the moon, the navigation chip was only 4K in size. Ten years ago, we thought that 20GB of data would never be filled, but back when the Manhattan Project was executed, the order of magnitude of its fission had already reached ZB. This is certainly not the first time humans have encountered the problem of big data. It is only the definition of "big" that is constantly changing.

So what exactly is big data? Research firm Gartner's concept of "big data" is a massive, rapidly growing, and diverse information asset that requires new processing models to have greater decision-making power, insight, and process optimization capabilities. In terms of data categories, "big data" refers to information that cannot be processed or analyzed using traditional processes or tools. It defines data sets that are beyond the normal processing range and size, and thus forces users to adopt unconventional processing methods.

From this definition, we understand that what is most important is not how big data should be defined, but how it should be used. Big data in the true sense can no longer be processed and analyzed by conventional processing methods, and the biggest challenge now lies in identifying the technologies that can better handle massive amounts of data and apply it to all aspects of life.

Alibaba's Wang Jian has some unique insights into big data, such as:

"Today's data is not big. What is truly interesting is that the data is now online. This is precisely the unique characteristic of the internet."

"For products in the non-internet era, their value must lie in their function. For products in today's internet era, their value must lie in data."

"Never attempt to improve a business using data, this is not big data. You must have done something that could not be done before."

Obviously, the real value of big data does not lie in solving the existing problems, but in creating and filling the countless gaps that have yet to be fulfilled.

So, what are the creative aspects of big data? Let us first take a look at two examples:

In the book *Big Data,* the author pointed out that big data has the power to transform public health. In 2009, the Influenza A H1N1 virus spread across the world in just a few weeks. Many people felt that the world was going to end. Both public health institutions and medical experts were pessimistic and believed that this flu would take many lives. What was worse was that the experts at that time had yet to develop a vaccine against this new type of influenza virus. The only thing that could be done was the slowing down of the spread of the flu, but the prerequisite to do so was to know where the flu virus appeared. Officials were helpless in this aspect.

It was at that time when Google made a statement. In summary, they said they could identify the precise locations of the flu virus, as they had all their users' search records accumulated over the years. Later, Google compared the top 50 million most frequently searched entries by Americans with the data from the US Centers for Disease Control and Prevention (CDC) during the period of seasonal influenza transmission between 2003 and 2008. The results found that the correlation between Google's predictions and the official data was as high as 97%. Like the CDC, Google could also determine where the flu started spreading. Moreover, their judgment was made very timely, unlike the CDC which could only do so one to two weeks after the flu outbreak.

Therefore, when the H1N1 flu outbreak occurred in 2009, Google was a more effective and a timelier indicator compared with the customary lag of official data.

Another example is in the Nepal earthquake that occurred in 2015. A website named HDX called for humanitarian data exchange. This website has a simple interface and tidy data classification, which can assist loose humanitarian organizations in working together. This website provided basic information about clinics, death reporting, and epidemic tracking functions during the Ebola virus outbreak in Africa. Currently, it mainly provides data from Nepal.

When the Philippines was swept by Typhoon Haiyan before the HDX website was available, Dale Kunce, a senior geospatial engineer for the American

Red Cross, asked for data to draw a rescue map route for first responders. The Philippines' government sent him over 40 pages of hastily scanned Excel tables. "They were not even scanned by a machine scanner. The papers were not in order, and the data was almost useless." As a result, dozens of volunteers had to be recruited for retyping the information on the spot.

Such backyard fire was not uncommon until the United Nations Office for the Coordination of Humanitarian Affairs and Frog, a global design company, jointly built the HDX website in 2014.

The support team for this website is made up of 2,000 emergency personnel from 80 countries. After analyzing millions of Nepali data, they built several large databases ranging from emergency calls to the reporting of specific emergency personnel who have arrived in Nepal, as well as reports about the earthquake types.

This data changed crises. The most important thing in earthquake-stricken areas is geospatial data. HDX can point out the locations of the road, the town, and the hospital. This way, rescuers coming from outside can be made aware of the locations that have been blocked and the areas where helicopters can land.

In the era of the mobile internet, data is being generated all the time. If this data is not used well, it will become a growing pile of garbage that will only occupy space. But on the contrary, if we can use this data in certain ways, it will bring convenience to our lives in various aspects, and even save many lives as per the two examples above.

Besides, big data will also play a pivotal role today, through the internet changes coming towards the medical industry.

5.1.2 Big Data Pre-diagnosis

In daily life, most patients usually choose to visit the hospital only when they are unwell. For example, female breast cancer patients usually seek medical treatment at the hospital when cancer has developed to the later stages and the patients feel severe discomfort. By then, they would have missed the optimal treatment window, and the treatment cost and difficulty that follow would increase dramatically.

But the above situation is likely to improve with health care that combines big data and wearable devices. For example, high-precision and sensitive sensors can be implanted in female underwear for real-time monitoring of various physical indices that may cause breast cancer. Timely feedback can then be relayed to the user's wearable device terminal. As such, the device will promptly warn users to visit the hospital quickly for diagnosis, conditioning, and treatment once the possibility of breast cancer is identified. For women, this is a need that is more pertinent than the so-called "rigid demand, pain point."

Therefore, one of the most important characteristics of the future of wearable health care is to move the current disease treatment model to disease prevention. The focus of the entire health care service should also be shifted from short-term acute disease treatment to chronic disease treatment and preventive health care. Moreover, through the integration of wearable devices and big data, cloud platforms, and artificial intelligence, it can also analyze, predict and judge users' daily behavioral habits and eating habits, and remind users to make corresponding adjustments to achieve the effect of disease prevention.

As more and more wearables enter our daily lives, the entire disease diagnosis and treatment process is also shifting from "treatment of disease" to "preventive treatment of disease." The realization of this scenario places high demands on the accuracy of sensors and the construction of big data analysis platforms.

First of all, in the early stages of data monitoring and collection, the sensor must ensure that the data is absolutely accurate, that is, it must be medical-grade. Secondly, after data is generated, the post-processing analysis of the data is crucial. This is the core of the entire wearable health care sector, and hence, the analysis requires the participation of professional medical teams. Finally, the storage and feedback of data involve privacy issues, which is also a threat that everyone in the internet era will be exposed to.

This application can realize the advancement of medical treatment, from "treatment of disease" to "preventive treatment of disease." At the same time, it can alleviate the problem of uneven distribution of medical resources. For more advanced applications, these wearable devices can be used directly as part of treatments, such as in the case of wearable defibrillators.

My Spiroo, smart hardware, can help patients evaluate when they may suffer from asthma attacks, as well as the external pathogens and factors which may trigger asthma attacks.

As My Spiroo has network capabilities, the development team planned to add geographic location information and statistical data to the paired application, as well as collect asthma data from users. This way, My Spiroo can send warning messages to patients who react to certain pollen or pollutant levels, which will assist them in managing their inhalers. Peter Bajtala said, "We will collect the data on the devices and visualize the data to understand the overall situation of asthma patients worldwide."

The medical system in the state of North Carolina state has begun using big data for the medical prevention of high-risk patients. Particularly for high-risk patients, a set of the algorithm was integrated into the user data of 2 million people to evaluate the incidence rate of diseases, to implement medical measures before patients fall ill. Take asthma patients, for example, hospitals can calculate the probability that a patient will be rushed to the emergency room by knowing whether he or she has had increased drug dosage, purchased cigarettes, and whether he or she lives in an area with high pollen concentration. Even for a gym member, the system can assess his or her chance of suffering from a heart attack by analyzing the type of food purchased.

The future of health care is as Gartner's analyst Robert H. Booz said, "The traditional rating and insurance industries have disappeared with the reform of health care, and now, efforts are directed towards active treatment management. We know you are at risk of diabetes, and we will take action before your symptoms show."

5.2   Personalized Remote Diagnosis and Treatment

In 1861, a French doctor rolled a thin notebook into a cylinder to clearly hear the sound of a patient's heartbeat, a move that gave birth to a stethoscope, which was a major step in clinical medicine. In 2016, doctors could view the electrocardiogram (ECG) of a patient on the other side of the globe while sitting in an office, through the information fed back by a smart wearable device. This is the medical technology revolution that is happening.

In the future, doctors and patients do not need to be in the same location, and they will still be able to solve all problems in the treatment process. I would call this telemedicine if I can be a little less precise. As people become more aware of their health, this market continues to be explored and is making progress year by year.

1. According to a report released by BBC Research and Towers Watson, the global telemedicine market would expand to USD 27 billion by 2016, of which virtual medical services would contribute USD 16 billion.
2. According to a report by IDC, 65% of the "interactions" organized by medical institutions would be done through mobile devices by 2018. By 2018, 70% of medical institutions would launch application software, providing wearable devices, and the will have remote monitoring capabilities providing virtual treatment.
3. In 2014, more than one-third of the funds invested by GV (formerly Google Ventures) went into the pockets of health care and life sciences companies.

In the health care industry, telemedicine will have profound implications for both patients and doctors. For example, remote consultation can reduce doctor visits by 93%, which saves consumers USD 103 per consultation and USD 1,067 per emergency consultation. According to the research report, 50% of doctor consultations can be done virtually, and 70% of electronic medical records can be completed through virtual means.

However, the biggest beneficiary of the prevalence of telemedicine is patients. With the emergence of new technologies, patients can receive faster, higher quality, and more effective treatments.

In the internet era, patients may still be required to sit in front of a computer to talk to doctors. The biggest drawback is that doctors cannot obtain the patients' medical data, so doctors are often unable to make accurate diagnoses. Ultimately, patients still cannot run away from the process of queuing for treatment in crowded hospitals.

But wearable devices will be able to solve this problem. Its greatest value lies in the big data collected from the human body and the use of this data for medical treatment. Particularly for the management of chronic diseases, sensors attached to various parts of the body will constantly monitor various physical indicators of the patient. When a patient feels unwell, the patient's doctor would have already obtained relevant data of the patient, including data that reflects his or her usual lifestyle. According to the data, doctors can quickly diagnose and provide suggestions for preventive measures and good living habits.

Let us look at two cases, one being how hospitals use telemedicine services to solve the problem of high hospital readmission rates of patients, especially for those with chronic diseases.

In 2012, the United States Centers for Medicare & Medicaid Services began to implement a punishment system for hospitals whose patients with congestive heart failure (CHF), chronic obstructive pulmonary disease (COPD), or pneumonia were admitted again within 30 days. In the next few years, the punishment system would cover more types of diseases, and the penalties would increase accordingly. Therefore, hospitals were trying all means to reduce the rate of readmission of patients within 30 days. This accounted for about 20% of all insured patients, but the readmission rate of patients diagnosed with the above three diseases was much higher.

How can this problem be solved? This can be done through a remote monitoring model, in which remote medical services are provided to patients, including remote home care, remote nursing, or telemedicine. Patients or their caregivers can provide daily updates of medical data directly from home, using wearable devices. The mobile tools currently used in most hospitals are computers, smartphones, or wireless communication devices. Data is transmitted to nurses or relevant medical personnel located in a medical service center, who flag up abnormalities for follow-up treatment.

Massachusetts General Hospital randomly arranged six months of routine post-discharge care or remote monitoring care for 150 recently admitted CHF patients for 6 months. The patients transmitted data of their vital signs to a nurse, who then reported any potential problems to a doctor. The readmission rate of patients due to various reasons and heart disease-related reasons in the remote monitoring group was lower than that of another control group (0.64 vs 0.73), although the difference was not statistically significant. However, this study was not intended to demonstrate a reduction in the readmission rate, but to assess the feasibility of remote monitoring of patients' conditions.

American futurist, Alvin Toffler predicted many years ago, "For future medical activities, doctors will face the computer and diagnose and treat patients based on various information about patients displayed on the screen, sent from afar." The traditional diagnosis method of looking, smelling, asking, and taking the pulse can be achieved through telemedicine. For example, with accurate sensors, doctors can diagnose the health of various physical signs of patients from the other end.

With the maturity and optimization of wearable device technology, sufficient to replace the original computers and smartphones, telemedicine services will become more efficient and effective. Particularly in terms of data processing, whether it is for the patient or doctor, all information will become clear at the premise of the establishment of a big data analysis platform. In this process, the role of doctors in diagnosing and treating patients will become weaker. They will focus more on the analysis of medical big data and corresponding medical research. In the future, patients will receive medical treatment and medical examination reports directly through the medical big data platform of wearable devices.

The concept of doctors and patients will be blurred in real life, as doctors will be virtualized, and patients will not be exposed to the public again and again due to frequent hospital visits, so they would not be constantly reminded that they are patients. Everyone will become a health expert and be equipped with an efficient private doctor. Particularly for China, which has a large population and scarce and unreasonable allocation of medical resources, telemedicine will be the key to solving these problems.

5.3 Health Management

5.3.1 Health Management

Health management first appeared in the United States. Although the United States has the world's richest medical and health resources, it could not afford the rapidly rising costs of health care. The pressures arising from the aging population and chronic diseases have cost the United States nearly two trillion dollars a year, but it has had little effect. The decline in production efficiency due to health problems has also threatened the development of the U.S. economy and society.

Relevant information shows that the unhealthiest 1% of the American population and the 19% who are chronically ill account for 70% of medical expenses, while the most healthy 70% of the population only accounted for 10% of the medical expenses. Everyone has a chance of becoming the unhealthiest 1% and the chronically ill 19% of the population. If attention is only paid to the ill population and investment is only made on "diagnosis and treatment" while the health risks to the healthy 80% of the population are ignored, the ill population will continue to expand and the existing medical system in the U.S. will be overwhelmed.

As the cost of medical services continues to soar, insurance companies and companies in the United States have noticed a shocking number: 80% of medical expenses were spent on the treatment of preventable diseases. Before World War I, the American Red Cross, Blue Shield, and other insurance companies found that medical insurance for large companies could be largely reduced if early disease prevention and management were carried out. Medical expenses could be reduced by 90% if 10% of health management costs were invested in the early stages. This is a large sum of money.

Past practice has proven that through health management plans, the incidence of diseases among American residents dropped significantly between 1978 and 1983, with hypertension and coronary heart disease falling by 4% and 16% respectively. Similarly, this secret formula applies to American companies and individuals, which is 90% and 10%. This means 90% of individuals and companies had their medical expenses reduced to 10% of the original amount with health management, while the 10% that did not participate in health management saw their medical expenses increase by 90%. Today, more than 70% of Americans in the United States are enjoying the services of health management organizations.

Presently, China is also faced with the issue of rising medical expenses that had occurred in the United States way back then. As standards of living continue to improve, the probability and risks of people falling ill increase due to various unhealthy lifestyles. This leads to increased health needs, resulting in heavier clinical medical burdens and increased medical expenses, aggravating the problem of it being "difficult and expensive to see a doctor" for residents. One way to alleviate this grim fact is to establish a health management system that focuses on prevention. The key steps include establishing a health record management platform and electronic case information. In short, these steps will be able to ride on the strong wave of Internet plus, let everyone participate in health management, and launch the true era of "preventive treatment."

5.3.2   Launching the Era of "Preventive Treatment"

Whether it is wearable health care, "Internet plus" health care, mobile health care, or smart health care, the ultimate goal is not to cure diseases, but to prevent the occurrence of diseases. This is also the basis of the entire big health era.

According to statistical research by the World Health Organization, 1/3 of diseases can be prevented in advance, 1/3 of diseases can achieve reasonable effects through early treatment, and 1/3 of diseases can have stronger medical effects through the communication of information.

In China's long-established culture of Traditional Chinese Medicine, there is the concept of "preventive treatment." Famous quotes like "the sage cures the ones who are not sick yet, not the sick," "a skilled doctor cures at the early sign of disease" and "protect the areas that have yet to be infected" continue to be circulated to date. Therefore, for the Chinese people, the traditional way of maintaining good health may have similar effects as "preventive treatment of diseases," though it may not be completely the same. The biggest difference is that the true era of "preventive treatment" is not built on the so-called diet therapy and health regimen, but on wearable health care, which accurately quantifies all vital signs.

Many people believe that more wearable devices will be used in the health care industry in the future. Today's wearable device applications are only at the initial stages of development, which consist of collecting data and sending it to the corresponding cloud platform. More in-depth applications will involve the process of big data analysis on the data collected, identification of the disease, and treatment. Only at this stage of application can it be possible to achieve the advancement of medical treatment, which is the transformation from "treatment of diseases" to "preventive treatment of diseases." Compared with the current medical conditions in China, the greater significance of this in-depth application is in its ability to alleviate the problem of uneven distribution of medical resources.

CHAPTER 6

Wearable Health Care Case Studies

On 14 February 1946, the world's first generation of an electronic computer named ENIAC was born at the University of Pennsylvania in the United States. It weighed 30 tons, took up an area of 150 square meters, and inside, it contained 18,800 electronic tubes. It was quite a huge monster. In 1981, the first personal computer was born. Since then, the personal computer has begun "flying into the homes of ordinary people." Who would have thought that slightly over 30 years later, today's electronic computer could be so light and portable and that it would continue to subvert people's lives?

Like the first-generation computers, many present medical devices are complex machines that are difficult for people without medical expertise to control. They often need to be used with various other devices to complete people's exams. On top of making occasional large errors, the devices may also cause mental and physical suffering to the people tested.

With the arrival of the wearable devices era, wearable health care devices that are miniaturized and convenient have relieved the agony of many patients during treatment. Moreover, the rise of wearable health care, digital health care, and remote health care has allowed patients to have the greater initiative in understanding their conditions during medical treatment.

We can look at the applications of wearable health care in real life in the following case examples.

6.1   Diabetes: Smart Diapers

Smart Diapers is a product developed by Pixie Scientific firm from the United States. It is a type of smart diaper. Besides having the functions of a normal diaper, the greatest feature of Smart Diapers is its ability to inspect and analyze a baby's urine to deduce the baby's health condition.

Parents or doctors of the baby can scan the QR code on the diaper through the QR code scanning function of the smart device to determine the health condition of the baby. The inspection's data can reflect many of the baby's health problems, such as urinary tract infection, long-term dehydration, and potential kidney problems.

According to a report from the World Health Organization, 10% of adults globally have diabetes. The scary thing is that this problem will continue to aggravate. Children or the youth are the most susceptible to Type 1 diabetes, and Smart Diapers can provide early intervention and treatment for Type 1 diabetes.

6.2   Cardiovascular Disease: ZIO Patch

According to statistics from the World Health Organization, three in ten deaths are caused by cardiovascular disease. According to Heart. org's report, cardiovascular disease has become the number one cause of death globally, totaling at 17.3 million deaths each year. This number is expected to increase to 23.6 million by 2030.

It is also worth noting that 80% of these deaths could be prevented. A wireless device launched recently called ZIO Patch can monitor a variety of arrhythmias such as atrial fibrillation, and its accuracy is even higher than Holter's Holter monitor (termed as Holter hereafter). For the past half-century, Holter has been a benchmark for mobile heart rate monitoring.

In actual experiments, Scripps Translational Science Institute (now Scripps Research Translational Institute) compared the ZIO Patches of 146 patients with the Holter electrocardiograph data (these patients' conditions were suitable for monitoring by both ZIO Patch and Holter). It was found that ZIO Patch identified a total of 96 patients with arrhythmia while Holter identified 61.

6.3 Assisting Doctors to Monitor Patients' Heart Health Remotely: CardioNet

CardioNet from the United States is a NASDAQ-listed and it is the industry's leading wireless medical technology company. The company focuses on the diagnosis and monitoring of patients with arrhythmia.

The wearable device MCOT developed by CardioNet can remotely capture and transmit patients' heart data. The device can complete ECG monitoring, analysis, and diagnosis in almost any location, thus ensuring that doctors can still obtain important medical information after their patients leave the hospital.

Joseph Capper, the President and CEO of CardioNet, pointed out that, "By getting patients to wear our new MCOT device, cardiologists can easily monitor patients and obtain accurate views of the heart's functions during daily activities."

Also, the data detected by MCOT will be transmitted to the company's monitoring center via the network. With the support of professional medical institutions and the company's research on big data, CardioNet's backend team can thoroughly analyze and diagnose this data and generate reports to send to patients. At the same time, when a patient is found to have an abnormal heart rate, the data can also be sent to the monitoring center for the timely analysis and recording by professional medical personnel, so that the doctor can be alerted when a life crisis is detected.

6.4   Health of Children and Pregnant Mothers: Ritmo

According to the World Health Organization, about 800 women worldwide die from complications related to pregnancy or childbirth every day. The wearable device Ritmo developed by the Nuvo Group allows expectant mothers to monitor the baby's heartbeat and movement 24 hours a day, while also assisting the mothers with providing early education for their children.

Ritmo is a safety band that wraps around the belly of a pregnant woman. There are three versions: Beats, Surround, and PregSense. Among them, Beats can detect the heartbeat and other physical signs of the mother and the baby due to be delivered. Surround builds on Beats by adding a prenatal education function, and it can play soothing music and other audio information to the baby through the mobile app. PregSense is the medical professional version of Beats. In addition to detecting pathological signs, it can also analyze and identify problems that occur during pregnancy and early symptoms of pregnancy complications.

The founder and CEO of Nuvo Group, Oren Oz, said that this device can record everything that happens during pregnancy from the surface of the abdomen and that their goal is to reduce the risk of relevant complications as much as possible.

6.5   What Happens if Medication is Not Taken: Proteus Digital Health

What happens if a patient does not take their medication? In a scientific article published in the *New England Journal of Medicine*, it says that "It is clear that only when a patient follows the doctor's instructions closely can effective medication take effect."

The article also states that 50% of patients will not take their medications as prescribed. But what if you can not only track whether a patient is taking medication, but also leave a lasting effect on the patient's physiology?

Proteus Digital Health is a medical services company. The medical services they provide are digitized, and their latest developed drugs are integrated with mobile, wearable, and cloud computing technologies to enable doctors and family members to be involved in the patients' management of health. Its core technology is a feedback system that allows patients to understand their health and their physiological reactions. A piece of disposable "medicinal plaster" patch attached to the skin allows the wearer to monitor his or her body's physiological reactions and behaviors. According to the Proteus website, the patch receives information from the sensor, monitors heart rate, activity and rest, and sends the information to the mobile device.

An article in the British *Financial Times* rated Proteus' plan as "a classic example of how a networked sensor can help medical professionals track treatment." At present, Proteus products have obtained EU CE certification and the U.S. FDA's approval. Furthermore, this company already has total financing of USD 291.5 million and has the potential and capital to expand the market.

6.6   Wearable Health Records: Google Glass

In June 2014, Drchrono, an electronic health record company based in Mountain View, California, developed a new application for Google Glass and called it the first "wearable health record". Doctors only need to register with Drchrono and they can download and use this app designed for Google Glass for free. They can then use it to record medical treatment or surgical procedures. Certainly, they will need to obtain the patient's consent in advance. The recorded videos, photos, and notes are stored in the patient's electronic medical records or Box's cloud, engaged in collaboration for cloud storage. The patient can refer to this information at any time.

A foreign startup company called Brain Power is currently developing a Google Glass app for children with autism. The program can play the most popular cartoon characters or pictures in Google Glass. When children with autism wear Google Glass to communicate with others, Google Glass can display different cartoon images to express the language and emotions on behalf of the children.

If the wearer wishes to make eye contact with the other party, he or she will only need to look at the other party's glasses. Google Glass will then automatically stop playing cartoon images and directly display the other party's face. The wearer can earn points for himself or herself in this way, just like the levels in video games that children are familiar with. Another Google Glass app developed by Brain Power can focus the wearer's attention on the speaker's eyes during a dialogue. This was created because research has found that children with autism pay more attention to the mouth of the speaker when communicating with others.

Google Glass can also enhance the sense of reality in surgical operations, provide doctors with full-body image information, and lower the rate of surgical errors. It can scan the QR codes on ward doors, medicines, and medical equipment to update and synchronize the correct patients' medical records. It can also take over most of the paperwork and recording work in the operating room. Google Glass's camera will continuously capture the doctor's actions in an operation, including the image and sound. It will then extract data from the video to help fill in information for electronic medical records.

According to MedCity News, an American website for health care news, Palomar Health (hereinafter referred to as the "Palomar Community") in San Diego County, California, is conducting a health care project called "Glassomics" (meaning "glasses economics"), which plans to use Google Glass for health care management, disease prediction and live video broadcast of surgical procedures.

Orlando Portale, Chief Innovation Officer (CIO) of the Palomar Community, believes that applications of Google Glass can be divided into six categories in the health care industry:

1. Differential diagnosis.
2. Ability to transmit videos of patients' conditions remotely.
3. Workflow applications in different departments, such as the emergency department accessing patients' medical records and receiving early warning information for patients in critical care.

4. View key symptoms, medical records, laboratory results, and other information.

5. Send operation procedures in video form for teaching or consultation purposes

6. Drug management, such as scanning of the bar code of drugs placed in the pharmacy or at the patient's bedside.

Many medical institutions and doctors are very optimistic about the future applications of Google Glass in the medical field, and doctors at individual hospitals have already begun to use Google Glass.

At present, six clinics in the U.S. are using the Google Glass software developed by Augmedix. When doctors are communicating with their patients, the software can automatically enter the patients' information into an electronic form. As Google Glass has a video function, the software can even understand the patient's non-verbal communication cues, such as identifying the painful area in the patient's body.

In the operating room of the University of California, San Francisco Medical Center, the attending physician, Pierre Theodore, used the display system on Google Glass to view X-rays without leaving the operating room or consulting colleagues from other departments. He could also control the device with voice commands.

Theodore pointed out that, "My eyes need to switch between the patient in front of me and the important information right before my eyes, but this does not distract me much. I think that Google Glass can be and will be a revolutionary product."

Besides, when Google Glass was first released, many people already had high hopes about its role in the medical field. They suggested that Google Glass could be used like this:

Radiologists can view body images without using a tablet or smartphone because Google Glass can directly display these images on the glasses worn by the doctor.

Surgeons can view image data while performing surgery, without having to look up when they need to view images or computers.

Patients can view any street and communicate with the medical service organizations in the area to understand the available medical resources, prices, doctors' rankings on Angie's list website, and the types of medical insurance they accept.

While maintaining eye contact with patients and colleagues, doctors can retrieve information about a certain disease.

Three years after the release of Google Glass, many of these once-conceived scenarios have now been truly realized, some even beyond imagination.

CHAPTER 7

Personal Wearable Health Devices

I t has been about three years since the concept of wearable devices was invented. During this time, wearable devices of various forms have been introduced in the market. They can cover almost all areas of the body and arm a person completely without leaving any gaps. However, with current levels of products' quality, if someone is truly dressed up as such, he or she will probably look like a garbage collector, dressed in garbage.

That aside, regardless of whether the current development of wearable devices as a whole has turned cold or will be further reignited, most of these devices are largely used for applications such as fitness and entertainment. Wearable devices in the medical field will remain highly popular while riding on the trend of "Internet Plus" (Figure 7-1).

Some people may confuse fitness wearables with health care wearables. Here, I will briefly clarify their differences. At present, most smart wristbands are fitness wearable devices, because their main function is to collect various physical data of users during exercise. Smartwatches can perform most of the functions in smart wristbands and smartphones, that is, they can be used for fitness, communication, and entertainment. On the other hand, wearable

Million USD

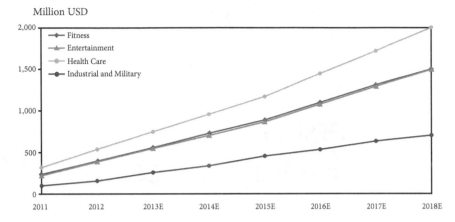

Figure 7-1 The development trend of wearable devices for different applications in the next three years

devices used in the medical industry have much higher requirements than those used for fitness. This is because such wearable devices require qualification certificates from medical institutions and doctors' approval. Each device may only have a single function, as it is developed to treat a particular disease. So, what devices are there? This chapter will introduce these health care wearable devices applicable from head to toe.

7.1 Head

7.1.1 Muse Brain Sensing Headband

The Muse headband was developed by InteraXon, a Canadian startup. It has seven built-in sensors that can monitor the brain waves responsible for users' speaking, critical thinking, and listening activities, and it can carry out corresponding operations after analyzing the brain wave signals. Its greatest feature is the combination of meditation mechanism with wearable technology, such that users can gain insight into the activities in their brains.

A neurofeedback application called Calm is used in conjunction with the Muse headband. This application simulates the user's brain waves through voices and images and shows them relatively peaceful and beautiful scenery, like a beach at sunset, helping the users to relax.

7.1.2   Halo Headband

In 2012, one of the founders of Halo discovered that when he came into contact with some bioelectricity in his basement, his brain experienced some wonderful feelings. From there, he began to study the relationship between stimulation and the brain. This is how the Halo headband came about. Moreover, many reports published in the past have said that certain levels of biological stimulation can improve the brain's ability to focus.

In 2014, Halo secured a USD 1.5 million seed funding from Marc Andreessen, a well-known Silicon Valley investor, who requested this high-tech headband be further developed into one which could stimulate the brain's potential through "neuromodulation."

Two of Halo's co-founders, Daniel Chao and Brett Wingeier, had worked for NeuroPace for several years. The latter created a brain-implanted medical device that supports the use of electrical stimulation to relieve the pain of patients with epileptic seizures. But Daniel wanted to create some head stimulation devices that do not require surgery and can be used by more people, besides patients. Moreover, recent studies by the University of Oxford have shown that cranial electrical stimulation can achieve considerable results even if not done internally.

The company had not disclosed many details, as it was applying for intellectual property rights and FDA approval.

7.1.3   CEFALY Smart Headband

The CEFALY smart headband is a smart wearable device developed mainly for migraine patients. It is also a Transcutaneous Electrical Nerve Stimulation (TENS) device that is specifically licensed for use before pain attacks. CEFALY is an alternative for migraine prevention medication. Christy Foreman, Director of the Office of Device Evaluation of the FDA's Center for Devices and Radiological Health, had said, "This device can help patients, who cannot take present migraine drugs, prevent or treat migraine attacks."

It is reported that this device is produced by Belgium's STX-Med company. In addition to the United States, it would also be sold in Canada, for USD 349.99 (about 2100 *Yuan*).

Migraines are characterized by strong pulses or throbbing pain in a certain part of the head, coupled with nausea or vomiting, and sensitivity to light and sound. A migraine can last from four to 72 hours if left untreated. According to the U.S. National Institutes of Health, these debilitating headaches affect approximately 10% of people worldwide, and migraines are more common among women, with the incidence three times that of men.

CEFALY is a small, portable, battery-powered prescription device. The user is required to use the self-adhesive electrode to position the device at the center of the forehead, directly above the eyes, with the two ends of the device placed on the ears. The self-adhesive electrode in the middle of the headband should be positioned between the eyes. The device then delivers an electrical current to the skin and subcutaneous tissue to stimulate the trigeminal nerve associated with migraine. The user will experience a stinging sensation or a massage feeling where the electrode is placed.

The FDA evaluated the safety and effectiveness of this device based on a clinical study conducted on 67 patients in Belgium. This study, which involved 67 patients, showed that patients who used CEFALY had significantly fewer migraine attacks per month and used less migraine medication than patients who used placebo devices.

7.2 Body

7.2.1 LUMOback Smart Belt

A smart belt called LUMOback launched by Lumo Bodytech may be able to improve the backache problem that plagues many people today.

LUMOback has a sensor embedded within it, which can be connected to the iPhone via Bluetooth. It can monitor the user's lower back. As soon as the wearer appears to be in a wrong sitting position, LUMOback will remind the user to correct his or her posture through mild vibrations.

In addition, LUMOback can track the user's daily activities, such as sitting, standing, walking, running, or sleeping. The accompanying application can also calculate the steps taken, calories burned, and duration spent standing, sitting, and getting up. The app would even rate users at the end of each day.

7.2.2 Hexoskin Smart Shirts

A startup called Hexoskin developed a type of smart shirt with the support of NASA and the Canadian Space Agency. During the day, Hexoskin can measure data such as heart rate, heart rate change or recovery, step count, calories burned, and breathing. At night, it can also monitor sleep and the environment, including sleeping posture, heartbeat, and breathing activity. All this data will be synchronized with the supporting application via Bluetooth, or uploaded online for real-time viewing by remote coaches.

This smart shirt is very suitable for users who love sports because its greatest feature is its ability to track the changes in the user's vital signs in real-time during exercise.

Just connect the phone, and one can see all the data it tracks. Real-time tracking requires users to carry their mobile phones with them during exercise, but if you usually use mobile phones to listen to songs, this is not a big issue. In addition to the application, users can view data on webpages online, and the data in both formats can be shared on social media platforms or handed over to coaches for further analysis.

At night, Hexoskin can monitor sleep quality, sleeping posture, heartbeat, and breathing activity. This is important data for athletes. This is because our body repairs itself and grows muscles at night. A healthy sleep cycle is almost as important as an effective training program.

7.3 Hands

7.3.1 Electronic Tattoo Thermometer

This is a novel wearable device developed by researchers at the University of Illinois. As its surface resembles a tattoo, the R&D team named it "electronic tattoo."

The device is an ultra-thin circuit that can be directly attached to the surface of the human body. It is mainly used to monitor the temperature of the human body to analyze the temperature peaks at different periods. It can also monitor the flow of heat in the blood through the veins, to understand the human blood circulation and the dilation and contraction of blood vessels. Professor Rogers, the project leader, said that this is very important for cardiovascular health.

At present, the main medical research institution in the United States, the National Institutes of Health has reached out to this research team. Roger said, "Currently, this temperature monitoring application on the skin is only at their first step. There are many areas in the human body whose monitored indicators can make great contributions to the medical field." The team is experimenting with ways to apply this device to the internal organs of the human body, going as far as placing the device on the atrium to measure temperature. They also published their R&D results in the Nature Materials journal.

7.3.2 Dialog

A San Francisco company called Artefact developed a wearable device to help patients with epilepsy. The Dialog they designed looks like a patch that can be connected to a smartphone, allowing patients to track, manage, and predict the occurrence of seizures.

Matthew Jordan, the Executive Creative Director at Artefact, recognized that the lives of people with epilepsy are very miserable. This group of people is often misunderstood by people around them and suffers from great anxiety.

Dialog connects with the smartphone application via Bluetooth. It can remind patients to take their medication in time, issue warnings during the onset of symptoms, and remind family members, friends, or caregivers. When the patient suffers from a seizure, the application can also tell the surrounding people what to do.

During a seizure, the device can sound an alarm as long as the patient is holding it. An epilepsy patient often experiences signs of seizure before it happens. At this time, he or she can double-tap the device to record it. This way, the patient can be prepared in advance and can also record the duration between the display of signs to the onset.

Matthew Jordan mentioned that this smart device was designed this way mainly to apply design thinking and innovation to complex medical problems, attempting to create a more humane experience, and help those chronic diseases patients who have difficulty seeking help.

7.4   Feet

"Why do we force consumers to wear personal items that never needed to be worn every day, instead of integrating sensor technology into everyday clothing?" This was what Davide Vigano, co-founder and CEO of Sensoria Fitness said in an interview with the media.

The ultimate form of wearable devices in the future must be invisible, but ubiquitous. Wearing them would be just as casual and natural as wearing clothes and shoes. We can understand how feet-hugging shoes became smart from the products below.

7.4.1   B-Shoe smart fall prevention shoes

According to statistics, on average, the elderly aged 72 and above will fall twice a year. Each accidental fall may have severe consequences. To prevent this from happening, an Israeli company launched a type of smart fall prevention shoes called B-Shoe.

From the outside, the shape of B-Shoe is not much different from ordinary shoes, but the secret lies in its sole. The greatest feature of B-Shoe lies in the presence of electric gears and sensors installed at the heel of the sole.

Once the B-Shoe detects via sensors that the wearer has lost his or her balance and is likely to fall, the motor will start to roll the shoe backward immediately, which will help the user regain balance.

7.4.2   OpenGo Smart Insole

The OpenGo smart insole was developed by Moticon, the creator of "The world's first fully integrated sensor insole." The beauty of this insole lies in its ability to turn our shoes into a wireless tracking system.

Shoes equipped with Moticon equipment can measure the pressure distribution and pressure change of the user's soles. The measurement data will be transmitted wirelessly and it can be viewed in real-time. Besides, the device is also equipped with wireless modules, flash memory, and a battery compartment.

There are 16 different types of sensors (previously 13 sensors) spread out across OpenGo, including three-dimensional acceleration sensors and temperature sensors. The insole also includes ANT+ wireless connection hardware and an integrated data logger. All the hardware is in a multi-layered structure. This includes the first layer which is the sensing layer and the second layer which is the dielectric layer, as well as the round electronic module at the bottom. The battery of the module can be replaced through a detachable lid on the sole.

The OpenGo insole can track the pressure on the foot and the gait during the acceleration of movement. Through the supporting application, Beaker, this information is directly transmitted to the computer as stored data for analysis, so that the impact on a user's feet during routine activities can be easily understood. This provides valuable data for athletes to assist them in recovering from injuries.

7.4.3   Smart Socks launched by Sensoria

Wearable devices have covered all aspects of the human body. Besides shoes and insoles for the feet, socks are also not overlooked. This smart socks launched by Sensoria, was unveiled as early as 2014, but it was only launched in 2015 at the CES 2015.

This pair of smart socks are not knitted with thread, but a special sensor fiber fabric. Due to its close-fitting nature, the socks can accurately detect relevant motion data, such as the number of steps taken during walking, the pace, and changes in running, walking, and even the body weight. The company also said that the collected data can create more products, and this type of fabric can also be used in wearable devices such as helmets.

Certainly, this pair of smart socks also needs an electronic sensor installed in the sock to collect corresponding data. After a single charge, this pair of smart socks can operate continuously for over six hours. Sensoria has also developed a supporting application for it, which is supported by the three major operating systems iOS, Android, and Windows phones, and can provide functions similar to voice coaching.

7.5   In vivo

7.5.1   Miniature Bladder Sensor

For patients with spinal trauma and those who are unable to control their defecation, the current solution is to insert a catheter into the patient's urethra and inject saline into the bladder. This treatment process is not only painful but also inaccurate. To address this issue, a Norwegian science and technology research institute developed a miniature bladder sensor that can help patients alleviate such pain.

This tiny sensor can be injected into the human body, and it can stay in the designated location for months, or up to a year. This way, patients will be able to sense the bladder pressure without experiencing the unbearable pain brought about by the catheters and saline. Moreover, this miniature sensor installed in the body allows the patient to move freely, which will reduce the chance of infection to a certain extent. It also has an external sensor that displays the bladder condition through a wireless connection with the sensor in the body.

7.5.2 Miniature Camera in Blood Vessels

It is common knowledge that the blood vessels in the human body are very small, so it is difficult for people to imagine putting a camera in such a small space. However, implantable wearable devices does just that.

This 1.4 mm ultra-micro silicon wafer can be placed inside the heart, coronary arteries, and peripheral blood vessels to capture 3D images in real-time. The results obtained by applying volume imaging on these images can help doctors better perform heart surgeries and avoid major operations through targeted dredging of blood vessels. To describe the device figuratively, this silicon chip is like a 3D flashlight placed inside the blood vessels.

This miniature camera imaging uses the CMOS technology widely used in digital cameras and smartphone cameras. The device uses ultrasonic transducers to process the signals on the equipment directly, then transmits data consisting of over 100 elements through 13 ultra-micro wires. Besides, the power consumption of the device is only 20 milliwatts. Low power consumption helps reduce the heating of the device in the human body. The miniaturization technology allows the chip to travel smoothly in the blood vessel and perform 3D imaging in the process.

Although there have been devices capable of capturing images inside of blood vessels before this one, those devices could only take cross-sectional images and could not provide doctors with the best view. Researchers from Georgia Institute of Technology said in a statement that the device allows doctors to observe everything inside the blood vessel, and it will help them to make a better judgment about the blockage within. The statement also said that the device may reduce the need for heart surgery by helping to unclog blood vessels.

7.5.3 PillCam COLON Capsule Endoscopy Device

The U.S. Food and Drug Administration (FDA) launched a PillCam COLON Capsule Endoscopy device, which is as large as a capsule and has a built-in miniature camera that can take color photos at a frequency of four or 35 frames per second. It can also support about 10 hours of wireless video transmission.

This miniature camera is aimed at colon polyps and tumors that are 6 mm in size. Before this invention, colonoscopes were long, thin cameras inserted through the rectum. There are 750,000 colonoscopies performed each year, and among them, women have a higher incidence rate as they undergo more pelvic surgeries and cesarean sections.

Besides, the previous model of colonoscopy needed to be supplemented with X-rays and CT scans during patient examinations, which made the process more complicated and more costly. Most importantly, this miniature camera device can exist very comfortably in the digestive tract and will not cause any discomfort to the patient.

CHAPTER 8

Business Models of Wearable Health Care

"Medical health" has gained more attention in the 21st century, and people are gradually gaining more awareness of it. By using this as an entry point to establish a business model in the health care industry, wearable devices are doing something subversive, highly meaningful, and valuable. Moreover, it is relatively easy to come up with a business model that conforms to the development pattern of the health care industry.

The biggest difference between the business model of wearable devices in the medical industry and other industries is that it does not make a profit purely through the sale of hardware.

8.1 Collaboration with Insurance Companies

According to a survey, Americans spend as much as USD 2.6 trillion on health care each year, and a considerable part of it is due to unhealthy lifestyle habits, such as poor eating habits leading to obesity and diabetes. The health care market has just begun to exponentially expand, and its potential market size is far larger than what could have imagined. I once explained in the book *Smart Wearables Changing The World – The Next Business Wave* that the combination of wearables and health care will lead us into the era of "preventive treatment."

According to my research on some business models based on wearable device applications in the United States, some insurance companies in the United States have begun integrating wearable devices into their industries, and have gradually formed unique business models that can be roughly classified as the following two models:

For the first type, the medical insurance company pays a portion of the service fees of the wearable devices for its insured users. For the other, the insurance company adjusts the corresponding insurance premium according to users' lifestyle habits to encourage users to adopt good lifestyle habits.

A typical example of the first model is Welldoc, a company that focuses on diabetes management. Its main model is its "mobile phone + diabetes management cloud." Currently, it focuses on mobile health care, but I believe there is a greater value that can be realized if it is combined with wearables.

At present, Welldoc mainly records and stores a patient's blood glucose data through their mobile phone, then uploads the data to the cloud. After analysis, it can provide patients with personalized feedback and send reminders to doctors and nurses. This system has proven its clinical effectiveness and economic value in clinical studies and it has obtained FDA approval for medical devices.

In addition, Welldoc's BlueStar application can provide services such as real-time updates, activity guidance, and disease education for patients who are diagnosed with Type 2 diabetes and require medication to control their condition.

Welldoc charges users accordingly based on the services provided. As the service provided by the company can help medical insurance companies reduce long-term expenses, two medical insurance companies have already begun paying over USD 100 per month for the "Diabetes Manager" system for insured diabetic patients.

The second business model is based on data mining and its use. Since most wearable devices have built-in sensors that can monitor and record various data that is closely related to human health at any time, insurance companies can use this data to understand whether the insured's lifestyle habits and various physical data are in good condition. They can establish a reward and punishment standards, in which premiums are reduced for people who consistently do sports and lead a healthy life and increased for those with unhealthy lifestyle habits.

Such an approach achieves a win-win situation. Through the analysis of lifestyle habits, insurance companies not only enable users to save on insurance premiums but also encourage users to establish good lifestyle habits. Besides, for insurance companies, the healthier the lives of their insured users, the lower the medical expenses they will incur.

In the United States, most medical insurance costs are currently jointly borne by companies and employees. Such an insurance method that combines the use of wearable devices can, to a certain extent, reduce the company's expenditure in this area. It can also encourage employees to exercise more and develop healthy lifestyles. This is killing two birds with one stone.

Evidently, the era of wearable devices changing the medical industry has already arrived. The big data it generates will not only bring about huge changes to the medical insurance industry but also to industries such as fitness and medicine. I have said many times at conferences, that a health and medical industry that is based on wearable devices will take the lead in establishing the business model. It will change and influence our lifestyles, as well as effectively improve the current health care situation.

8.2 Social Wearable Health Care

8.2.1 The Initial Formation of the Wearable Health Care Community

In recent years, with the advent of the mobile internet era, social media platforms have developed rapidly. These include Twitter, YouTube, and Facebook overseas, and Weibo, WeChat, and Zhihu in China. Social media platforms have fundamentally reconstructed the way people connect, build communities, and share information.

Facebook has 1.3 billion users, and Twitter has over 900 million users. Meanwhile, for YouTube, an hour-long worth of videos are being uploaded every second. It has reached 396 million monthly active users in over countries and regions worldwide. Socializing in the mobile internet era has shifted to online platforms. This further breaks the barriers between countries and regions, and truly enables the entire world to be connected in an instant.

These social media platforms drew groups of people from all walks of life, and later formed small social circles of different characteristics according to geographical regions, interests, occupations, and other factors. For example, there are groups of patients who wish to inquire about their illnesses, and groups of doctors who wish to get more medical research cases through social media platforms. In other words, medical communities have gradually formed and they continue to exert influence on social media platforms.

Emily F Peters, the founder of Uncommon Bold, believes that "health care is a perfect match for the social media platform." Uncommon Bold has an internet medical background, and it has participated in the operations of startups four times and has worked for Practice Fusion and Doximity. Not long ago, she wrote an article describing nine new ways to use social media in the medical industry. They are showcasing the operating room to the public, crowdsourcing medical diagnosis challenges, raising millions of dollars' worth of funds for clinical research. They are also using #FOMO to raise public health awareness, uncovering life-saving information from social media data, directly selecting subjects for clinical trials, providing a communication platform for doctors, making organ donation a new trend, and awakening individuals' attention to special diseases.

Mobile socialization of medical and health services will certainly be the direction for the next major development because anything can happen wherever the crowd gathers. As health care is closely related to life, it has always been a particularly cautious and strictly regulated industry. This often results in a habitual lag in its marketing and communications. But today, many doctors have discovered the groundbreaking changes that social media has brought to health care, so they have begun to spare no effort in using various forces to promote the development of social media in the medical industry.

For example, the nine cases of social media complementing health care as mentioned above are being implemented in some places abroad. It was mentioned in Uncommon Bold's article that to satisfy some medical enthusiasts' curiosity about surgery, the Swedish Medical Center in Washington State, US, broadcast a live cochlear implant surgery, and later showed touching scenes of the patient listening to music for the first time. Memorial Hermann-Texas Medical Center in Texas broadcast the cesarean section of a six-pound baby boy live on Twitter. UCLA also live broadcast of a Parkinson's patient playing country music on the guitar during his brain surgery.

In 2014, the doctors participating in World Vasectomy Day live broadcast 25 vasectomy operations performed in one day. Over 10,000 viewers watched the operation process, international video interviews, and short documentaries online. All these publicity was done to popularize some common knowledge and alleviate people's unnecessary fear of surgery.

8.2.2 Domestic Medical Social Ecosystem

The most commonly discussed topics about present wearable products are that they are not fashionable enough, their battery lives are terribly short and the user experience is poor, resulting in poor customer loyalty. However, we overlooked another aspect, which is the social ecosystem based around wearable devices. This is a new opportunity for most terminal mobile portals in the mobile internet era.

In this era of social media, where everyone posts content, is eager to party round the clock and subscribe to one another's content – if the wearable device ecosystem can be built based on socializing, this may give Chinese companies a differentiation advantage compared to international giants.

On 1 July 2014, four wearable devices were launched on JD.com at the same time. They were smart wristbands from iHealth, Huawei HONOR, Lifesense, and Codoon. In the field of wearable devices, where creative new products emerge every day, product forms like smart wristbands can be considered "antiques." Presently, no product is particularly brilliant in design either.

If the product is merely a smart wristband with no social media mechanism established or integrated within it, it is not worth a mention. Of course, it is considerably more difficult to build a wearable device-based social ecosystem than to launch hardware, but it is an opportunity.

I have a thought – we must use the current instant messaging tools to explore and try them. I can make products such as the MOMO version of a smart wristband, the WeChat version of a smart wristband, the Weibo version of a smart wristband, and the Laiwang version of a smart wristband.

Take the instant messaging tool WeChat for example, a "WeChat version of a smart wristband" is not a Tencent smart wristband. In short, it is a secondary development through the WeChat port which uses WeChat to manage the data in the wristband for instant sharing, viewing of fitness rankings of one's WeChat friends, and interaction, challenges, and competitions among friends.

All four wristbands launched by iHealth, Huawei HONOR, Lifesense, and Codoon have this function. They can use WeChat to synchronize and manage the data from different brands of wristbands through official accounts, and link this data with WeChat's social circles for users to share with their friends on their Moments timeline and to view "fitness rankings." I approve of such initiative.

Smart sports wristband is the most common product form in the field of wearable devices currently, but these wristbands are made by different manufacturers and have no unified system platform. Hence, the user ecosystem is fragmented. Different brands of wristbands usually require the installation of their unique apps, and data cannot be transferred between them. Besides, industry standards are still in the exploratory phase, and each brand has its own way of defining parameters. The user base of each product is no different, no one is particularly prominent with attributes that could allow users to make their choices without hesitation based on the unique advantages.

From a psychological perspective, sharing accomplishments with one's closest social circle brings about the strongest sense of satisfaction. Satisfying users' joy of sharing is also a way to improve customer loyalty, to make up for some deficiencies in the technological development of the hardware itself.

Judging from the current situation of the wearables industry in China, for hardware companies, in particular, the apps that were created based on the respective hardware do not look promising and have become white elephants for users, possibly due to their fixed models. I have been observing this phenomenon. Many users do not post the accomplishments from their wearable devices onto their respective apps, but take screenshots to share on WeChat Moments or Weibo instead.

Perhaps, drawing lessons from this phenomenon, traditional hardware companies still need to learn from WeChat to establish their social ecosystems. They can also start with the current approach, which is to leverage mature social media platforms, such as Weibo and WeChat.

As far as manufacturers are concerned, connecting their products to WeChat at this stage is a matter of course. This is because the companies will no longer need to invest money and effort to develop the corresponding apps. Moreover, when a product has integrated social media functions, boring health data can reach a wider range of people. It can also stimulate user activity, similar to game rankings. All of these factors will improve customer loyalty ultimately.

In the future, the establishment of a social ecosystem for wearables, the new entry point for mobile terminals, will be a new business opportunity for Chinese companies.

8.2.3 Three Major Models for Entry to Health Care Social Networking

Industry giants like BAT enter the market directly through users, hospitals, or doctors, and they form a strong closed-loop advantage. It is relatively difficult for entrepreneurs to seize opportunities in these areas. Perhaps another market segment that has not received much attention yet could be a good choice for entrepreneurs who wish to invest in the "Internet Plus" medical industry is social networks based on the medical industry chain. There are generally a few categories:

1. For patients: This means a social network that is created for a specific patient population, such as liver disease patients, diabetes patients, heart disease patients, and kidney disease patients. These patients with the same disease can share their treatment process, treatment results, and be prescribed

medication on the social networking platform. They can also share about the service levels of the respective hospitals they were admitted to and the quality of treatment they received from the medical staff. This can provide an intuitive reference for people who contracted the disease and are looking for medical solutions. This will be a new opportunity for "Internet Plus" health care.

2. For medical staff: In the early days, the use of social media in the medical profession was restricted by strict privacy regulations. But now, the medical industry is quickly adapting to the new situation. In addition to career advancement, the biggest firm need of medical staff is an improvement in diagnosis and treatment levels. Therefore, social networks that target medical staff per these specific characteristics could pose new opportunities. They include users of specialized fields like surgery, internal medicine, hepatology, and pediatrics. These users can share papers and research findings through such specific social networking platforms. They can also discuss and share the diagnosis and treatment methods of their respective cases.

    Professional websites such as Sermo and Doximity provide doctors with dedicated socializing and collaboration platforms. Doximity is a pioneer in the field of information collection for medical groups. It provides its doctor members with corresponding medical training and information such as the income of doctors from all over the world. The Mayo Clinic Social Media Network is also a leader in this regard. It has been committed to helping doctors and medical institutions get involved in social media.

3. For regional personnel: This is another type of subdivision, separated according to geographical regions. An example would be a social network for patients in a county, where they can discuss and share medical diagnosis and treatment experiences in or outside the region. The advantage of such social interaction lies in the cohesion of the local community.

8.3 Exploration of Business Models

8.3.1 Crowd Segmentation

A research report on the market for wearables, released by the German consumer research company, Growth from Knowledge (GFK) in October 2014 stated that one-third of wearable devices were "discarded" by users within six months of purchase. The American magazine, *WIRED*, also published an article which said that more than half of the fitness tracking device users in the U.S. no longer use their various devices. One-third of these users kept the devices in their drawers or gave them to friends after using them for less than six months.

Wearable devices have poor user loyalty. There are many reasons for this. It may be because the price is too high, the design looks too poor, or the functions are too weird, among others. However, I think the most fundamental problem is that the functions offered by the devices are not the firm needs of users. In other words, the devices have not identified the users' pain point.

Heart rate monitoring, step count, and other functions on the devices could be used as gimmicks in the early stages, but smart users soon discovered that this data was not very accurate. Moreover, users used to lead equally enriching lives without these monitoring devices in the past. This is why many products aimed at health management, fitness health care, and social entertainment did not do well. They were forgotten too easily.

Wearable devices certainly cannot continue to be what they set out to be, that is a product which is like a jack of all trades. Instead, they have to perform vertical segmentation of the market and develop specific functions for different groups of people. Particularly in the wearable health care industry, it is only through sufficiently precise segmentation that products can truly impress users, occupy the market, and integrate into the lives of users. Therefore, ways to orientate towards the market, segment it sufficiently precisely, truly impress users, and occupy the market will be the main directions of future development for the wearable health care market.

According to the principle of market segmentation, wearable devices can draw up their market plans according to different product forms such as fitness wristbands, smartwatches, and smart glasses. This will clarify their goals.

Furthermore, it is more important to segment the market according to groups of people, such as infants, children, women, the elderly, and the disabled, and develop devices tailored to their needs.

1. Devices for Infants and Toddlers

Infants and toddlers have poor awareness in all aspects and they have special needs, such as constant monitoring. Hence, wearable devices designed for them will have stricter safety requirements.

The main functions of such wearable devices are to record and monitor health indicators such as sleep quality, turning motions, body temperature, and heart rate of infants and toddlers. It is also to transmit the data to parents' computers or mobile phones and perform certain processing and analyses. Also, if there is an unexpected situation where a baby crawls out or falls out of the cot, such devices will need to send an alarm to the guardian immediately through text message or other means.

Recently, a variety of wearable devices developed for infants and toddlers have been launched to help young parents become masters of infant care quickly.

A smart foot band for babies launched by Sproutling can perform real-time monitoring of a baby's motion and heart rate, and indoor environment (including temperature, humidity, noise level, and lighting). It also has a positioning function that is used to match the weather data of the city where one is located.

This foot band is composed of three parts: indoor sensor, wrist strap, and mobile phone software. The band is made of white medical material and has a cute red heart in the middle, a built-in battery, and four sensors. Also, Sproutling established a special database of health data for babies aged zero to one years old. Parents can input their babies' age, weight, height, and other data in advance, then connect to the device using their mobile phones. When an abnormal data analysis is detected, the Sproutling foot band will sound an alarm immediately and automatically draw the attention of the parents. This way, even if parents are not by their babies' sides, they can watch over their babies' conditions at all times. They will thus be able to sleep well, without the fear of them sleeping too heavily or unfortunate incidents happening, such as their babies falling out of the cot.

At present, similar products on the market include smart socks developed by Owlet Baby Care, smart baby pajamas produced by Mimo and Exmovere, and smart diapers developed by Pixie Scientific in New York. China does not have self-developed wearable devices for infants and toddlers at the present, but the company Babytree recently launched B-smart, a smartwatch specifically for pregnant women. This device runs on Android and can monitor pregnant women's weight and fetal movement, and record exercises and contractions.

The complexity of accurately monitoring all aspects of a fetus still in the mother's womb is much greater than that of a baby that is born, but this is another market with development potential. Babytree also planned to continue launching a series of devices to monitor information related to baby growth, but the specific plan had not been disclosed.

Be it a fetus in the mother's womb or an infant, both belong to a very special group. If a wearable device can accurately track, record, and analyze all aspects of data under very safe conditions, its market potential will be infinite, and it will likely become a firm need.

Besides, wearable device companies can also connect with young parents, a group that is closely related to infants and toddlers. Companies can target them to train them to quickly become masters of parenting. This can be done by providing them with step-by-step courses that are a combination of theory and practice. The product form can also be upgraded according to babies' growth needs. This can improve user loyalty to a certain extent. The product may even be able to accompany the child through the entire childhood.

All in all, the four main requirements of wearable devices for infants and toddlers are safety, comfort, accuracy, and timeliness. If companies wish to enter this market, they will need to build a complete service platform in addition to the hardware. In particular, companies that produce infant products traditionally must leverage their understanding of this market and the user groups accumulated in the past to launch wearable devices at the right time. They will stand to gain more easily with the first-mover advantage in this market.

2. Products for Children

According to a study, about 200,000 children go missing in China every year, and only 0.1% of them are eventually found. Behind every lost or missing child is a family that can no longer feel complete. Child safety has also gradually become a more highly regarded public safety issue in society.

At present, most wearable devices for children in the market have relatively simple functions, which are mainly positioning and tracking. This is also one major "battlefield" for wearable devices which is tailing behind health management. Most wearable devices that are used for ensuring the safety of children are based on the "hardware + software + cloud" three-in-one operation mode. In addition to developing basic hardware, companies also develop corresponding mobile phone applications and data analysis platforms. This not only provides users with a more comprehensive and high-quality experience but also opens up more avenues for potential profits.

At the end of October 2013, Qihoo 360 launched a "child safety wristband" priced at RMB 199. The device could monitor the safety of children in real-time and it had functions such as safe area warning and call connection. FiLIP, an overseas brand of a smartwatch for children can use GPS, WiFi, and cellular data to send a child's location information to his or her parent's mobile phone, as well as make calls.

In China, Eachpal launched a SmartUFO watch especially for children. The biggest difference between this and other devices is that it has an additional WiFi positioning system. By detecting the MAC addresses of WiFi hotspots in the surrounding environment, SmartUFO can determine the specific latitude and longitude of these hotspots, then calculate the location of the device. The device's positioning accuracy is up to 20–100 meters, and its power consumption is only one-twentieth of GPS. It can last up to two weeks in between charges.

Numerous wearable devices have been entering the market for child safety since 2013. In addition to the ones introduced in the text, there are also hereO, a children's smartwatch which targets users aged three to twelve, Tinitell, a wrist-type children's mobile phone, LeapBand, a child activity tracker, and LG's wearable device named KizON. A common feature among these devices is the positioning system, but they also have a common shortcoming, that is, the devices lose their functionality when separated from the children. For instance, children may unconsciously take off the device, drop it, or lose it. Human traffickers may also remove the device deliberately and discard it, which will fail to position the child.

In addition to displaying children's positioning, wearable devices with stronger battery life, and more precise geo-positioning, companies or entrepreneurial teams that wish to enter this field will need to find ways to solve the critical problem of the device being separated from the child for various reasons.

3. Obese People

In May 2014, international health researchers did a study on the global obesity situation and gathered that nearly 30% of the global population were obese or overweight, and a total of 2.1 billion people were obese. This problem poses a heavy burden for both poor and wealthy countries.

In addition, China has 46 million obese people, ranking second in the world. Recently, Baidu's Vice President Zeng Liang pointed out that based on Baidu's big data analysis, the most searched term on Baidu by the 290-million strong Chinese female netizen population was "weight loss."

As social material living standards improve and food-related problems surface, the age group of obese people has been lowered. Over the past 30 years, the ratio of boys and girls who are obese and overweight in developed countries has reached 24% and 23% respectively, while the ratio has reached 13% in developing countries. This proportion continues to rise.

In 1980, the number of overweight and obese people in the world was 857 million. By 2013, it had increased to 2.1 billion. The number of obese people in the future will continue to increase with the growth of the global population.

The World Health Organization recently stated that approximately 3.4 million adults die each year from various chronic diseases such as cardiovascular disease, cancer, diabetes, and arthritis caused by obesity. Obesity has gone from being a matter of body shape to putting a person's life at stake.

However, no country has yet to come up with a good strategy to deal with this problem, nor a solution that can truly reduce its country's obesity rate.

The obesity problem has become an important public health challenge on a global scale.

To tackle obesity, dieting or liposuction are unsustainable strategies that do not solve the root cause of the problem. Only regular exercise and diet, and long-term good living habits can fundamentally solve this problem.

Therefore, I believe the true "pain point" of the weight loss and fitness market will be about finding ways to make this group of people willingly exercise, to help them achieve weight loss naturally, establish good lifestyle habits and improve their physical fitness index.

Wearable devices that became popular in recent years have become a stepping stone into this market. In the future, it will become the most competitive product in the weight loss market because of its obvious advantages, mainly in the following four aspects:

First of all, wearable devices can be worn on the body 24 hours a day. At present, no smart device can do this. Even when using mobile phones, users will try to shut it down and place it far away when going to sleep at night, due to its radiation.

Secondly, wearable devices can monitor user health data in real-time, round the clock. This is the greatest value of wearable devices at present because the data generated can be used in all aspects of life, and it can create an impact particularly in health care, where it can lead us into the era of "preventive treatment."

Thirdly, the social media function of wearable devices and their collaborations with medical insurance companies will encourage users to continue to participate in sports and establish good lifestyle habits.

In addition, the market is currently experiencing high-growth activity for sports and health devices or mobile phone applications. As the advantages of wearable devices in this area gradually surface, their high popularity in the fitness and weight loss market will follow.

I was first to say that the health care industry will become the growth area for the wearable device market. According to the market segmentation criteria, wearable devices can also invest efforts in the group of people who wish to lose weight.

It is evident from the above description of the market background and advantages of wearable devices, that the wearable devices will be able to accomplish a great deal in the weight loss market. Therefore, I suggest investors and entrepreneurs of wearable devices to consider entering the market via the market segment of weight loss. This will be another market with great potential for wearable devices.

4. Products for the Elderly

Elderly care has become a common problem globally, especially with chronic diseases and nursing problems that come with old age. They have become a large health care financial expenditure in many countries. To tackle these two problems, wearable device developers can develop devices that are suitable for the elderly to use at home, and they can build corresponding community care and big data platforms to alleviate the problems of elderly care.

Particularly for China, where the concept of elderly care is highly regarded, smart elderly home care will become the main future trend. According to relevant surveys, 90% of the elderly choose to stay at home for elderly care, and only about 10% choose to stay in centers. Within this background, the key driver for the market will be ways that can extend elderly care services for the elderly who stay at home, as well as to meet their needs for socialized elderly care services.

In addition, the number of patients with Alzheimer's disease (commonly known as dementia) is increasing year on year. At present, more than 30 million elderly people worldwide suffer from Alzheimer's disease, and more than a quarter of them are in China. Alzheimer's disease patients suffer from a partial loss of ability to take care of themselves. The most prominent problem is their inability to find their way when they go out, so they would get lost if they are relatively far from home. Therefore, in addition to recording health indicators such as heart rate and respiratory rate, there is also a need for wearable devices in the elderly market to provide real-time positioning of the elderly.

Such products are also presently in the preliminary development stage. There are no outstanding products on the market as yet. The main product types are shoes, mobile phones, or accessories. The Comfort Zone product

CMA800BK recommended by the American Alzheimer Association is the size of a business card and weighs 50 grams. After putting the product in the patient's pocket, guardians can obtain information on the patient's precise position in real-time with just one click. When the patient walks out of the Comfort Zone, the guardian will receive a text message notification immediately.

This product is based on Qualcomm's inGeo platform, which requires a network connection, has a built-in GPS chip, and it can be remotely controlled. The product was sold for USD 99.99, but there was an additional monthly service fee of USD 14.99. Also, if the elderly can carry a smartphone, they can directly use Sprint's smartphone to use these functions through a dedicated Comfort Zone. An additional monthly service fee of USD 9.99 would be required for the service.

Many Alzheimer's disease patients like to walk around, and they are somewhat stubborn when introduced to unfamiliar products, so they may take a while to get used to using products like mobile phones and sensors. Positioning shoes jointly developed by the American GTX and Aetrex shoe companies managed to make the tracking equipment invisible, so the elderly would not even realize that they are carrying a GPS device.

This shoe product looks no different from ordinary shoes, but with the built-in GPS chip, which guardians can use to locate patients in real-time through mobile phones and computer software. The product also has a notification function based on the safe zone.

5. People with Disabilities

Ways to help the disabled lead better lives and integrate into society have been on the minds of people who care about the disabled across the world. Wearable devices are undoubtedly the best way to help the disabled. There is a huge market and room for the development of a series of wearable devices to help the disabled communities live like abled people. Examples include using Exoskeleton to help paralyzed people stand up again, using special glasses to help the blind regain "sight," and using advanced equipment and systems to help the deaf-mute "speak" again. Many technology companies have embarked on the development work, or better still entered the market.

(1) Artificial eyeball

A biological startup company developed an artificial eyeball, which uses the EYE (Enhance Your Eye) system to allow the blind to see the world again. The company mainly uses 3D printing technology to create human organs. Thus far, they have successfully manufactured ears, blood vessels, kidneys, and so on. However, according to the person-in-charge, it is rather difficult to print an eyeball successfully, due to its complexity.

The company currently offers three different styles of eyeball systems. EYE HEAL is the standard version with electronic iris, EYE ENHANCE includes additional electronic retina and filters (vintage, black and white mode etc.), while the highest-end EYE ADVANCE even supports Wi-Fi communication. It is as if the eyeball has become an electronic device like a mobile phone...

To use the EYE, patients will need to remove their original eyeballs, then install the Deck retina to "match" the brain.

The researchers said that artificial eyeballs are projected to be launched only in 2027, and they have no pictures of the relevant physical product for the time being.

(2) Anxiety control device for children with autism

Children with autism may sometimes have great difficulty speaking, especially when they are anxious. Hence, teachers and parents need to be cautious, so as not to make these children anxious.

Research shows that about half of the children with autism have gotten lost on their way to school from home, or back home from school. This is partly attributed to their anxiety. Another reason is that they were caught off guard in the face of danger.

To tackle this problem, the wearable device field has launched two devices that can control the uneasy emotions of children with autism, namely Neumitra and Affectiva. These two devices are designed to measure human physiological responses. They can be used for various medical purposes, such as tracking information like patients' post-traumatic nervousness and anxiety. These smart wristbands can also serve tens of thousands of people with autism, making it easier for their caregivers to track their anxiety levels.

At present, relevant organizations have begun testing the Affectiva wristband. According to the Autism Society of Ohio in the United States, school teachers would distribute these wristbands in class, and the teachers would then use these wristbands to monitor the behaviors and actions of the students. That is when the students are most relaxed.

(3) Consciousness control wheelchair

A startup company called Emotiv held a design competition, in which it invited developers to use the company's neurotechnology to create new relevant applications. One of the applications allows people with limited mobility to control wheelchairs through their consciousness.

Emotiv has developed a headset that can receive electronic signals from the human brain and convert these signals into actions. Redg Snodgrass said, "A wheelchair controlled by the human brain can make it easier for people with disabilities to achieve their goals, and relevant devices will 100 percent be the future for wearables."

The idea of a consciousness-controlled wheelchair started with Albert Wong, a Malaysian law graduate who suffered from Duchenne Muscular Dystrophy. Wong's family contacted Emotiv and asked the company to produce a system that could allow Wong to better communicate with others through a combination of mental commands, facial expressions, and head movement. Although Wong passed away shortly after, Emotiv planned to continue working closely with the disabled, especially those who are paralyzed from the neck down.

(4) Emotiv Insight Brainware

Emotiv Insight Brainware is a wearable brain wave tracking device jointly developed by Philips and information technology consulting giant Accenture, to provide Amyotrophic Lateral Sclerosis (ALS) patients with the possibility of controlling their surrounding environments.

Emotiv Insight Brainware can scan a patient's EEG brain waves and draw a "brain-computer interface" accordingly. The data collected by the device will then be transmitted to the tablet computer, allowing the ALS patient to issue commands through the tablet computer to control Philips' electronic appliances, such as smart TVs, light bulbs, or even make calls for medical emergency services.

(5) Motion Savvy

Motion Savvy is an innovation company supported by Leap Motion's hardware accelerator AXLR8R. Unlike other companies, all nine current members of the company, including the founder Ryan Hait-Campbell, are deaf. The company's CEO and founder Ryan Hait-Campbell was deaf at the age of five and learned to speak with the help of a machine from a young age.

Ryan Hait-Campbell hoped to develop a low-cost but effective device through his efforts to help more deaf-mute people communicate with the world like him. A few years ago, he teamed up with a software engineer, Alex Opalka, and later brought in Jordan Stemper and Wade Kellard to co-found Motion Savvy.

The Motion Savvy product released by the company is called "UNI." It includes both hardware and software. The hardware portion is a tablet device about the size of an iPad mini, with Leap Motion sensors embedded. When the (deaf-mute) user wants to speak, Leap Motion will capture the sign language, and Motion Savvy's software will interpret and translate the sign language first into text displayed on the screen, then "verbalize it" through a human interpreter's voice.

Thus far, Motion Savvy has not been officially launched, and it only supports American Sign Language for the time being. Motion Savvy has been "entering" sign language information for the system in the past year. Ryan Hait-Campbell planned to make Motion Savvy a system similar to Google Translate, with one end recording sign language information produced by the deaf-mute community and the other directly translating sign language into speech.

6. Professional sports products

Unlike wearable devices for public health, professional sports smart devices need to measure athletes' vitals like heart rates and respiratory rates more accurately, and monitor data like their speed, running distance, and endurance on the field. In the later stages, more professional data analysis kits are needed to help team doctors understand the different physical conditions of each athlete, to develop an individualized training and recovery plans. Also, coaches will need to understand the state of the players more intuitively and select the most suitable players to play.

This type of product has not attracted the attention of the general public for a long time. At present, the main product types are jerseys, sports underwear, and wearable devices especially developed for sailing, mountaineering, golf, boxing, and other sports. The biggest difference between this type of product and other ordinary wearable devices lies in its ability to formulate specialized sports training plans for athletes, accurately point out athletes' errors during training, and provide targeted guidance and corrections.

Not long ago, Zlatan Ibrahimović, the Swedish football star of the Paris Saint-Germain Football Club, took off his jersey after the game and revealed black underwear similar to a woman's bra, which piqued the curiosity of fans. This was sports data underwear launched by professional sports equipment company GPSports, which monitored players' physical and sporting conditions on the playing field in real-time.

Besides Paris Saint-Germain F.C., the Spanish Real Madrid C.F., the English Chelsea F.C., and many other European giants have also been using GPSports' products. It is just that the athletes rarely revealed their "bras."

In addition to GPSports' sports data underwear, there is also a product named Bro, which directly embeds a recording chip into the rugby jersey. This is also a GPS recording chip that can help team coaches and team doctors understand each players' condition. If a player's state of physical fitness and activity decline, the coach can get an intuitive answer through data analysis on an iPad or computer, then choose another player in a better condition to play.

8.4   Leveraging Wearable Devices to Reduce Medical Costs

International Data Corporation (IDC) data shows that in 2013, more than 6 million medical and fitness equipment was sold, and this number would rise to 100 million in 2018. Considering this was at an early stage, and the number is expected to continue to soar.

Karen Taylor, Research Director of the Center for Health Solutions at Deloitte Consulting, said, "Increasingly, patients are even asking doctors for recommendations for wearable devices."

Wearables are being used in different situations to assist patients with Parkinson's disease, diabetes, heart disease, high blood pressure, and other diseases to manage their diseases, as well as to help the elderly live longer and live alone.

According to statistics published by the British PA Consulting Group, this technology reduces hospitalization rate and consultation rate, saving 6.5 billion pounds for the United Kingdom National Health Service every year. Therefore, the United Kingdom made huge investments in the research of telemedicine and is considered to be a representative of the earliest application in this area.

"It is foreseeable that wearable devices will generally reduce overall medical costs in the future." Simon Segars, CEO of the British company ARM, said that people in remote areas can transmit high-definition data and receive remote analysis and treatment at home. This eliminates the pain of traveling.

Wearable devices, when combined with the internet, big data platforms, cloud computing, and professional physicians will simplify the entire medical treatment process. For instance, if you have high blood pressure or heart disease, when you are about to reach the limit for drinking, the wearable device will trigger a warning to stop you from drinking further and will suggest appropriate food you should consume to take care of your body.

Suppose you are sick but do not know what disease you are suffering from, traditionally, you may consider going to a cumbersome hospital. But during this time, countless doctors are monitoring your condition through the small wearable device, which will receive a physical examination report and prescription in a very short amount of time. Moreover, the prescription would have been sent to a partner pharmacy, and within the next 10 minutes, your medication would be delivered to your doorstep.

In short, future medical treatment will reduce overall medical costs to a large extent, especially the cost of patients' time. This is precisely the most prominent problem of traditional medical treatment at the present. In the future, everyone will be able to understand their own physical health at any time and become the manager of his or her health, while doctors would only play an assisting role.

In addition, wearable devices can also reduce the cost of research and development of new drugs. One of the biggest problems faced by current drug developers is the high failure rate of new drugs in the clinical development process.

Data from the Tufts Center for the Study of Drug Development shows that the average cost spent on developing a new drug from the research and development stage to FDA (Food and Drug Administration) approval is as high as USD 2.5 billion. Of which, the relatively high drug failure rate has largely contributed to the increase in R&D costs. When the cost of drug development is high, its cost will naturally and ultimately be passed on to the patient. Therefore, the ultimate beneficiary of a reduction in the cost of drug research and development is still the patient.

The question now revolves around ways to reduce costs. One can use wearable devices to collect sufficient daily behavioral data and clinical data about patients, build on this to create a more effective drug development process, and prioritize resources for the most effective treatment. This is because big data analysis and new clinical technologies, such as mobile health solutions and wearable devices, can greatly change the way clinical trials are performed. They can increase the value of data and obtain more results from clinical trials.

The analysis and processing of big data are not only done with higher speed but also greater accuracy. When doctors ask how patients feel, they often get subjective answers. Subjective data is also important in scientific research, but it pales in comparison with objective data. Both of these factors will allow more effective drugs to be developed at a faster rate.

In addition, wearable devices can track and collect a series of new objective data in real-time, allowing doctors to evaluate the physiological indicators of patients in real-time, and understand how much a drug affects the quality of patients' life. This is an increasingly important measurement indicator for pharmaceutical companies, regulatory agencies, and insurance companies.

For example, 23andMe, a Silicon Valley-based genetic data company, appointed Richard Scheller, the Executive Vice President of Research and Early Development at Genentech, a genetic engineering technology company, as Chief Scientific Officer (CSO) and Head of Therapeutics. He was responsible for mining nearly 900,000 people's genetic database information for 23andMe, with hopes of finding new clues for the treatment of common and rare diseases. Some companies have also joined forces to study ways to better deploy wearable devices for patients, combining data from these devices with traditional clinical data to measure changes in patients' behaviors and to later use this information to make more informed decisions.

CHAPTER 9

The Bottleneck of Wearable Health Care Development

9.1 Invalid Data

Most people have the general perception that the larger the amount of data, the better it is. Particularly in the medical industry, a lot of quantified data of pathological signs is indeed needed to assist doctors in diagnoses, but the premise is that the data is accurate, true, and effective.

9.1.1 False Data

The standardization of mobile health care big data analysis is not transparent enough, and its disorder can affect the judgment of practitioners. This is because the "big data" that most people see nowadays is information that has been integrated, analyzed, and processed, and few people have access to the original raw data. Any research result that lacks original data is questionable, as we do not know how the information was processed through the information integration process. Just like when the fruit is processed into juice, we have no way of knowing whether the fruit was rotten or fresh before processing.

Similarly for the predecessor of readily available "big data," or "raw data" in other words, the ultimate users of the data have no way of finding out what type of data it was – whether it was relevant garbage data or relevant target data – before the data was processed. Conceivably, applying such data into practice for clinical care guidance would be a terrible idea.

Take the "Google Flu Trend" of the past as a simple example. It is an ironclad case example, which can prove that people's fears are not unfounded. Google has a tool called "Google Flu Trends," which tracks internet users' search keywords and other related data to determine the flu situation in the United States (for example, patients will search for the word "flu"). The working principle of this tool roughly works like this: the designer inputs some keywords (such as thermometer, flu symptoms, muscle pain, chest tightness, etc.), and when a user enters these keywords, the system will begin its tracking analysis and create a regional flu chart and a flu map.

However, one will find that the actual application of this data is much more complicated than the theoretical program. First of all, the results of Google searches are supposed to be related to the "self-assessed" influenza incidence of the user population, but when the results of the study were published, the report itself had an impact on people's normal lifestyles and behaviors, due to Google's huge influence. Not only did it cause inconvenience to people's lives, and made a mountain out of a molehill, many people were also affected by it through their search. They would habitually or unintentionally pay attention to this data, causing much "noise" to be created in big data, which affected the big data itself, and ultimately led to inaccuracies in big data analysis.

Moreover, too many random and low incidence events are included in the generation of mobile health care big data. Take for example a family doctor who gives a patient an electronic sphygmomanometer and instructs the patient to measure and record his or her blood pressure every day. All of us find ourselves in different types of environments every day, and blood pressure will certainly vary according to the situation.

For instance, the patient could be holding a birthday party today and could have won 5 million from the lottery yesterday. One would be in high spirits on happy occasions. They then work overtime the night after and watch a sad movie two days later. Their blood pressure will certainly change because this is the body's self-regulation mechanism at work. This has no reference significance for medical dynamic monitoring, also because it is impossible for doctors to be fully aware of the various situations in the patient's daily life. The concept of a rise in blood pressure here has been surreptitiously changed, and the conditions are not valid, either.

Certainly, there is still a long way to go for the maturity of health care big data. Immature big data is not only unable to provide us with the basis for accurate reference and judgment, but it may also mislead and interfere with our judgment, thereby affecting the development and decision-making of the entire industry. In this way, it is natural for us to have lingering fears. After all, there is nothing more tragic than a half-baked big data project.

9.1.2   Inaccurate Data

At present, the hardware type that is most closely related to "Internet Plus" medical treatment is smart wearable devices. They are distributed in various parts of the human body to monitor the physiological signs of the human body in real-time, and they continuously generate a large amount of data, such as heart rate, blood pressure, and calorie consumption. However, the key question is whether the data collected by wearable devices can truly be used as a reference for doctors during diagnosis and treatment?

Jawbone UP is a smart wristband that can track a user's daily activities, sleep and diet. It can even monitor the duration of the user's deep sleep, and this data is then fed back to the user through a mobile app.

However, the data obtained by the Jawbone UP wristband is not all standardized and accurate. For example, on the function of monitoring the user's duration of deep sleep, Jawbone UP determines whether the user is in the state of a deep sleep by tracking whether the user moved, as well as the magnitude of movement during sleep. It records this data and relays the feedback to the user.

Medical scientists explained that the judgment of the depth of sleep is based on the performance and characteristics of electroencephalogram (EEG), electromyography (EMG) and electrooculogram (EOG) during sleep. Sleep is divided into different types. At present, besides the deep sleep monitoring method based on EEG, which is used in conjunction with EMG, EOG and electrocardiogram (ECG), there are no other scientific and mature monitoring methods, yet. Jawbone UP's method of judging whether the user is in deep sleep by the range of body movement is very unscientific.

Recently, scientists from the University of Pennsylvania invited 14 volunteers to take part in a test, which used wearable devices and mobile phones to perform 500-step and 1500-step tests. It was found that the average error of step count detected by the three apps installed on the mobile phone was about 6%. On the other hand, the error on the wearable device was very large. The error even miscalculated the number by discounting it by less than 22.7%.

The health care industry is one that has to be more rigorous and prudent than any others. Therefore, the accuracy and effectiveness of data are crucial. However, many small teams who are worrying about their survival cannot carry out large-scale, reliability verification. To save costs, many products turn out to be inherently incomplete.

Wearable devices are considered to be one of the main forces driving the transformation of the health care industry, and the most prominent aspect of this force is its data value. However, if products with health data tracking functions cannot collect data in a more standardized and scientific manner, it will be difficult for them to be recognized by the medical community. Inaccurate data would even mislead users and result in erroneous analyses.

In addition to inaccurate data, current mobile health care products also lack systematic data mining. Much of the physiological data is simply fed back to smartphone applications, without processing and interpretation happening in the background. Many users cannot even understand the data. Such situations drastically reduce the value of the data.

Many mobile health care products were aimed at "health management," but they ended up as gimmicks to fool users. Most products are only capable of monitoring and recording simple data. They lack in-depth data mining and have poor interaction. Some health management applications can provide only very little information, which is insufficient for diagnosis and treatment advice, and can only be used for a general health assessment. for a general health assessment.

Besides, due to the lack of large-scale data processing, multi-dimensional data analysis, and in-depth data mining capabilities, even if the collected data contains a lot of useful information, including data that can be used to diagnose diseases directly, it will be lost in the sea of data due to the inability of processing and minning the data.

9.1.3   Those Irrelevant Mobile Health Care Platforms

In August 2014, Aetna, the third-largest insurance company in the United States, declared that it would stop investing in CarePass, a mobile health data platform. It shut down CarePass mobile and network services before the end of the year. The project was aborted in less than two years.

At the beginning of the project, Aetna was confident that users would upload their data to the CarePass platform using wearable devices and that they would set health management plans based on their health goals. In addition, users could look up disease information and make appointments for medical treatments through the CarePass platform.

CarePass was an open platform that allowed users to pair and access multiple wearable devices, but its biggest flaw was that users could not make anything out of CarePass' consolidated data. For example, they could have uploaded their running distance, heartbeat, heart rate, calories burned, and duration of sleep. But this data was scattered. Users could see the changes in the data, but they did not know what the data meant as a whole.

The average user does not have professional medical knowledge, and data that has not been processed and interpreted will be meaningless to them, let alone useful. Take a diabetic patient, for example. CarePass may be able to integrate his regular blood glucose data, as well as his walking distance and the number of calories burned every day, but the user may not find any relation between his health and the data. Even if he thinks there is a correlation, he does not know what it is.

It is normal for users to not understand the data. And if a mobile health care platform is merely used for data collection, it is worthless to users. Sooner or later, they will lose interest. To engage and retain users, the platforms need to consolidate and analyze the data, generate reports or explanations that users can understand, or even make reasonable and effective recommendations. When users feel that their health conditions have indeed improved, they will naturally develop product loyalty.

There are many reasons for data invalidity, such as immature technology, irresponsible individual enterprises, data fragmentation due to division between enterprises, and so on, but there is a more critical factor, that is, the scarcity of data analytics talent. This is a common challenge in the whole data analysis industry. The technical requirement threshold for big data analytics is usually higher. Moreover, analysts need to understand the business, analyze, and program... In certain cases, they may even be required to do some system platform maintenance. There are very few of such all-round talents.

Big data is like a double-edged sword for wearable devices. Without an effective big data analytics feedback platform, data loses its fundamental value, which will in turn cost wearable devices their core value. On the other hand, effective use of the data allows wearable devices to achieve success quickly, making them the best hardware in the mobile health care industry.

9.2   Privacy Security

The International Data Corporation (IDC) predicted that by 2020, the total amount of big data in the world would be 57 times the sum of all sand grains on all beaches on Earth, and this amount would double every two years. But the sheer volume of big data also poses many challenges, one of which is the protection of personal privacy.

In the health care industry, large amounts of medical data are generated every day. One type of data is the objective record of the medical process. More importantly, mining these objective data can assist doctors in clinical decision-making. But during medical data collection, processing, and application, data leakage occurs from time to time, which leads to disclosures of patients' private data. Even anonymity or the protection of important fields cannot ensure the security of personal privacy. It is still easy to identify specific individuals through the collection of other information.

The medical industry can be said to be "closely related" to people, having as close as a "life-and-death" relationship. All medical activities, including the collection and application of health care big data, revolve around people during this process and are built on people's information. Hence, to ensure the security of health care big data, we need to make a judgment from both the "people" and "data."

First, let us take a look at the security of the "people" in the security of health care big data. It deals primarily with data privacy issues, including the personal privacy data of doctors and patients.

A mother who has just given birth receives a series of promotional calls. Before the operation, she had already been approached by a fraudster posing as the "operating surgeon" to solicit red packets... The security risk of big data goes without saying. Although many people may assume the personal privacy of patients to be personal medical information, the two are certainly not equivalent. Medical treatment exists for understanding, intervening, and restoring the human body, it is an organ-related organism. The collection, storage, transmission, and processing of information for understanding the state of a patient's nervous, circulatory, respiratory, digestive, and other physiological systems, such as blood pressure, pulse, heart rate, and respiration are personal private data from the perspective of social ethics.

In the medical treatment process, the privacy of patients mainly includes individual physical characteristics, health conditions, interpersonal contact, genetic fund, medical history, and medical records involved in a health check-up, diagnosis, treatment, disease control, and medical research. From the perspective of privacy owners, patient privacy can be divided into two types: the first type is personal information that people do not want to be exposed, which is related to the specific individual and whether he or she has confirmed the details. Examples include an identification number and medical records. The other type is common information that certain groups of people do not want to be exposed, which is related to the confirmation of sub-specific group mechanisms, such as the distribution of a certain infectious disease. Knowing the similarities and differences between patient privacy and personal medical information can help to clarify medical information and the subject of its privacy protection. This will aid in the distinction of medical data that is private and requires heavy protection, and medical data that can be shared and used.

Compared with the privacy of patients, the protection of doctors' privacy in real life may often be overlooked. This may sow the seeds of hidden dangers in medical development and the treatment of doctor-patient relations. We cannot ignore the personal privacy of doctors that is "sacrificed" for reasons like "responsibility" and "compliance with professional standards." This includes and is not limited to: being able to identify individuals, or being able to express specific religious beliefs, political preferences, criminal records, and sexual orientation.

Certainly, to help patients, medical service providers, health management agencies and insurance agencies understand doctors and their medical activities, technical processes can be deployed to conceal or obscure identities, and the private information in personal sensitive data can be deleted before analysis. This includes the understanding of the doctor's treatment habits from the diagnosis.

Secondly, as a form of "data," medical big data also has considerably large security problems. There are two main potential hidden dangers, one being a significant target of cyberattacks. In cyberspace, there is a high degree of attention on health care big data. The sensitive data it contains attracts potential attackers. The second danger is a challenge to existing storage or security precautions, especially when complex and diverse data is stored together during consolidation. Conventional security scanning methods will not be able to meet its security requirements.

Regarding privacy security, it is necessary to first tackle it from the national legislative level, then spread the information through media reports to raise the public's attention regarding information security issues.

In 2014, the National Health and Family Planning Commission published the *Measures for the Administration of Population Health Information (for Trial Implementation)*, which pointed out that privacy protection requirements must be met, the security of population health information must be maintained and private information cannot be disclosed during the collection, management, and utilization of population health information. But these are only sporadic mentions of a few articles. There are no specific laws and regulations about the protection of patient privacy in the context of big data applications as of yet.

During the Two Sessions in 2015, Wang Jingcheng, a deputy to the National People's Congress, put forth the proposal that relevant laws and regulations must first be improved in protecting patient privacy. For example, with regard to the ownership of electronic diagnosis and treatment files, it is necessary to clarify that while the patient's electronic diagnosis and treatment information is generated in the hospital and is collected and kept by the hospital, it is still unclear whether the owner of the information is the patient.

In North America, patient rights and health information are protected by law. In 1996, the U.S. Congress passed the Health Insurance Portability and Accountability Act (HIPPA), which protects patients' privacy with legal means, in a bid to plug all loopholes. The U.S. Food and Drug Administration (FDA) recently confirmed the mobile health care network security supervision and set the guidelines for the network security of medical devices being introduced to the market, especially the guide to GxP cloud hosting security. It requires medical devices' risk assessment documents to be provided. Risk assessment includes identifying equipment, threats, and vulnerabilities, and assessing the impact of these threats and vulnerabilities on the device functions and the users/patients.

It is reported that in the EU, wearable device manufacturers and suppliers must consider privacy and security issues to comply with the law. These must be handled properly in the design phase, the data capture phase, and the early phase of data collection.

The EU's data protection law stipulates that individuals must be informed in detail about the data collection of themselves, such as what data is collected, how it is used, and the individuals' rights to manage this data.

In China, the Measures for the Administration of Population Health Information (for Trial Implementation) (hereinafter referred to as the "Measures") formulated by the National Health and Family Planning

Commission stipulates that information shall not be collected out of scope, private information shall not be leaked, and responsible individuals will be held accountable for violations of these measures.

Certainly, detailed rules will need to be continuously added to protect the data security of people to the greatest extent, and to keep up with the entire ecosystem of the mobile internet in the future.

9.3 Gaps in Top-Level Design

After the Two Sessions in 2015, "Internet Plus" became a trend, and "Internet Plus" medical treatment was the most sought after. It saw countless related mergers and acquisitions, repeated investments, and cross-industry developments. The "Internet Plus" medical industry has already highlighted the lack of supporting policies and regulations, industry access and standards, and industrial planning in the current momentum of development that is exceeding expectations.

While many problems remain in the traditional health care industry, new problems will arise with the continuous development of internet health care. For instance, ways to protect the interests of patients and the type of standards to set up may create new forms of doctor-patient relationships, and also bring about new problems for supervision.

In addition, for telemedicine consultation, there are problems in the definition of legal responsibilities and medical responsibilities between the patients and the platform, and patients and doctors. There is also a lack of industry standards for internet medical treatment, and the interconnection of information between various institutions has not yet been realized. It is also difficult to break through the traditional chain of benefits. For example, online medical services such as *chunyuyisheng* is in opposition to hospitals. This is because online medical service is encouraging doctors to practice more medicine, but hospitals want doctors to stay in the hospital. All these issues will pose higher demands on supervision.

Recently, Song Shuli, a spokesperson for the National Health and Family

Planning Commission, said that other than telemedicine provided by medical institutions, other activities involving medical diagnosis and treatment are not allowed on the internet. Only health consultations are allowed. This speech was regarded by the industry as an urgent adjustment to the booming internet health care industry.

Perhaps, the "top-level" design of relevant government departments can start from the following three aspects to provide a basis for the current development of the wearable health care industry and a clear direction for its future development.

Strengthening policy guidance

At present, there is a lack of particularly authoritative interpretation information of the industry, such as the relevant data on the scale of the internet industry and the distribution of the industry, the total amount of existing funds, and the industry layout, among other questions. These are mostly analyses made by industry insiders and calculations by relevant institutions. The data may not be completely true, and the analyses may not be objective enough. Therefore, the government has a great advantage in this regard. It can fully invest the highest quality resources in the research of the entire industry, carry out corresponding statistics and analyses, consolidate the requirements for the healthy development of the industry, and publish relevant reports promptly to guide the steady investment of various funds and keep the location, industrial structure and industrial layout reasonable and scientific.

Improve relevant laws and regulations

For China, wearable health care is still in its infancy, but with the rise of "Internet Plus" in 2015, the development of wearable medical care was bound to accelerate. This is because it was at the core of the integration of Internet Plus health care. But at present, there is a lack of relevant laws and regulations in this industry, and this will become one of the factors that hinders the development of this industry.

In other words, appropriate policies will promote the development of this industry and protect the rights of all parties. As the saying goes, without rules and standards, nothing can be accomplished. In the absence of policies, doctors, patients, hospitals, or investors will merely be following one another and not making breakthroughs. For example, some developers do not know what kind of supervision the programs and devices they develop will be subjected to. They will spend their time on other pursuits, without caring about what regulators will do in the future.

While learning from the industry regulations in developed countries and studying other domestic regulations of the internet industry, the problem of lagging regulations in the internet health care industry should be solved as soon as possible. This will ensure that in terms of market access, legitimate competition, industry standards, service scope, service models, definition of rights and responsibilities, consumer rights protection, corporate obligations, and technical requirements, the internet health care industry has laws to follow, and the laws must be followed. By regarding the legal system first and governing through the law, fairness, justice and sustainable development of the industry can be maintained.

Strengthen supervision
While improving management laws and regulations, it is necessary to strengthen the management of industry access, rebuilding, expansion, change, and other permits, and put an end to market disruption through the failure of admittance where needed, and the failure to review where required, and failure to record where necessary. It is also important to strengthen the supervision and management of investment and financing funds for enterprises, technical systems, management and quality personnel qualifications, enterprise registration status, service scope, and service platforms, to prevent unqualified, unlicensed, and unauthorized enterprises from providing services.

Lastly, it is also necessary to strengthen the supervision of corporate services to the society, to the public, and for diagnosis and treatment, to prevent the occurrence of disputes without proof, service behavior without evidence, and dispute resolution with no basis. This will ensure that the development of the industry is always under supervision.

CHAPTER 10

The Prospects of Wearable Health Care Development

10.1 Mining the Value of Big Data with "Openness"

The greatest value of wearable devices lies in big data, but without the validity and sustainability of the data, the value of wearable devices will be drastically reduced.

Wearable devices are a product that relies on the internet and was developed during the mobile internet era. However, in the later development stages, it lost the spirit of the internet completely. In other words, we are applying the closed-loop thinking of traditional industries to an innovative industry that is blazing new trails. This is a problem that needs to be looked at.

What is the spirit of the internet? In my opinion, one of its most important concepts is "openness." This is essential for the development of the wearables industry. Why do I say so? It is common knowledge that all that the core value of wearables lies in data, but presently, no one dares to engage in the wearables industry in the so-called "free" way. I can say with certainty that presently, if anyone has the guts to provide free hardware and use the wearable devices to collect data, to realize its value indirectly, they will be treading on a path headed towards a dead end.

The only likely difference would be the way one traverses this pitted path. If you are a listed company, the victim will be the investors. If you are a company backed by an investment organization, the victim will be the investment organization. If you are a buddy-style entrepreneur, the victims will be the buddies. What can this pitted path offer consumers? The upside is an extra free technology toy, while the downside would be leaving users in utter disappointment.

If the wearables industry wishes to realize the value of data mining relatively quickly, it needs to have the spirit of the internet, which is "openness."

The prerequisite for openness is to have companies support the platforms required for openness. To catalyze the realization of the value of data, the industry needs to build on this to promote two aspects of "openness." The first is for platform operators to have an open mind so that different access platforms or wearable device companies can reap the benefits of data mining on the platform, based on the value of their collected data.

The second is for wearable device companies to have an open mind and be willing to collaborate, team up and share data, so that data can be processed by professional platforms in a standardized manner, as opposed to individual companies holding on to small amounts of data but dreaming of great value.

One of the critical reasons why the commercial value of data from wearable devices tends to be zero is that we are still using the closed-loop thinking of traditional industries to manage "smart wearables," a product with the openness concept. If the industry continues to remain closed, holds inconsistent data, and continues developing in silos, can this path lead to anywhere? I cannot say no for sure, but in my opinion, this would at least stretch the development of the wearable device industry even longer, and there would be an even more spectacular show of martyrs following in one another's steps.

Although there are many societies, alliances, and organizations in the wearables industry, I think the most realistic and sensible thing to do at the moment is to explore ways to centralize the data of respective organizations and the members in the organizations. This will be more meaningful than discussing the so-called standards. This is also a major business opportunity. The value created in the platform built is far greater than the current hardware itself.

For wearable health care, this link is particularly important and it is the most valuable to the groups of people on the two ends of wearable devices. However, the *2014 Smart Wearables Industry White Paper* published at the end of 2014 showed that the churn rate of wearable devices within three months was as high as 87%. There was poor user loyalty, and difficulties in maintaining continued usage were then the biggest bottleneck for the development of health care wearable devices. The biggest reason for this situation was that the data collected by smart hardware devices of different brands were not uniform and the sample sizes were insufficient. At present, true big data analysis cannot yet be performed.

Therefore, the top priority for the health care wearable industry should be to build an open and effective big data platform together quickly.

10.2 Integration of "Terminal" + "Cloud" + "Service"

In 2014, among smart wearable devices, wristbands and watches continued to thrive in the industry, and they first seized a sizeable market share. But beneath the glorious performance were layers of threats, such as high levels of product homogeneity, poor user experience, lack of aesthetic design, and so on. As a result, many smart wristbands have even shown signs of regression in the market.

This is an inevitable process for the entire development process of wearable devices. A market blinded by its roaring business requires appropriate cooling measures to get back on track and focus on the critical points for the development of the industry. For instance, the biggest problem which wearable devices currently have is in their finding a way to build an open platform that can serve the people at both ends of the devices, which will end the current exploration phase where everyone works in silos.

This core super-platform requires three key factors, namely the terminal, cloud, and service factors.

The magazine *Wired* once wrote that the elderly, chronic disease patients, and people with low income were the biggest beneficiaries and they were the groups with "firm" demand for wearable devices. Only when wearable devices become a necessity, not a dispensable health product in people's lives, only then they won't be easily abandoned.

According to a report published by Hong Kong's *Wen Wei Po*, the École Polytechnique Fédérale de Lausanne (EPFL) launched a new type of lens with an amazing function. Users can enlarge images with just a blink of an eye. This lens is expected to improve the vision of 285 million people with low vision around the world. This lens will become indispensable for people with low vision in their everyday life, travel, and work.

As the core medium of the whole mobile medical industry, wearable devices can record various data of the human body, but their capabilities have remained stagnant at the "recording" function. Even the data "recorded" requires further verification. Ideally, the wearable device on the user's end would be capable of accurately recording data of the body condition, transferring the data to the cloud for storage in a point-to-point manner, then processing and analyzing the data with a set of scientific algorithms established in the early stages. And finally, concluding feedback to the user and their doctor, fitness coach, or guardian.

Prior to falling ill, the use of wearable health care can help users establish healthy living habits and accumulate necessary health tips. During sickness, it can plan the entire medical treatment process for patients, provided that they are no longer conducted in hospitals like in traditional medical treatment, but can be done at home. Upon recovery, wearable devices can continue to monitor the users' recovery progress, with emphasis on certain areas. For instance, for cancer patients, wearable devices can understand the condition of cancer cells and predict the possibility of recurrence, and so on.

But to achieve all these aims, wearable devices alone are certainly insufficient. So, what is key? The aim is to integrate wearable devices, clouds, and services. Currently, the data from wearable devices is fragmented and system platforms are working in silos. These heighten the importance of integrating devices, clouds, and services.

There is still much room for improvement in the accuracy of data recorded by wearable devices. Furthermore, breakthroughs are required in terms of design and user experience. Simply put, users will not invest their time to a device that is neither aesthetically pleasing nor user-friendly.

As for cloud computing, a special "health care cloud" needs to be created. Building on new technologies such as cloud computing, the Internet of Things, and multimedia, the health care cloud should combine medical technologies, apply the concept of "cloud computing" to build a medical health service platform, and use cloud computing technology to build a new health service system. This will achieve the purposes of improving the service efficiency of medical institutions, reducing service costs, and facilitating residents' medical treatments.

The greatest value of a health care cloud lies in the providence of different application solutions for community residents, medical service personnel, and medical and health management personnel. For example, residents can use wearable devices to automatically upload personal physiological data collected onto a dedicated data center for remote monitoring. The remote monitoring system will automatically push the results of each data analysis to any type of smart device owned by family members, to keep them updated about the residents' conditions.

In addition, alarms for the detection of abnormal data can also be set in the system. When the collected data is found to exceed the healthy range, a warning will be triggered for medical staff to provide timely advice. This way, doctors can monitor and advise on health conditions without meeting patients face-to-face. Residents can also obtain their health information conveniently and quickly. At the same time, medical staff can improve their daily work efficiency and better monitor the entire treatment process.

The "service" here refers to the various services provided by relevant medical staff, especially for patients using telemedicine. Peripheral services such as payment services for patients during the consultation process, medication delivery services, and fitness programs tailored for users during routine exercise will give rise to many business opportunities. As the medical industry is a strict one, it is still difficult for ordinary companies to meet its requirements of entry, but they can provide many peripheral services. Moreover, services of different nature will ultimately support the entire service platform. This is also a critical element for wearable health care.

The integration of "terminal" + "cloud" + "service" is not only applicable to the wearable health care industry but also a development goal set for the entire wearable device industry. The key to this development lies in the Internet of Things. If wearable devices can be integrated with more services and applications such as smart homes, smart cities, and smart health care, there will be far more room for development and imagination.

10.3   Precision Health Care

When Apple's CEO, Steve Jobs, was suffering from cancer, he spent USD 100,000 to sequence his tumor and his entire genetic profile. Famous Hollywood star Angelina Jolie also revealed to the media that she opted for breast removal surgery after having performed genetic testing. This reduced her risk of breast cancer from 87% to 5%. The powerful celebrity effect brought about a surge in genetic testing and has made this technology a trend.

Genes contain all the patterns that dictate the full cycle of life from birth to death. Precision health care brings about huge transformations to the health care industry. The core concept of precision medicine is to subdivide the population, treat and diagnose the subdivisions accurately, and to have highly accurate interpretations of the unique behavior and data of the subdivisions, to provide accurate and tailored solutions.

In the future, genetic testing will become an essential part of medical treatment. More and more people will learn about their vital signs and health conditions through genetic sequencing.

Bina Technologies, a startup established in 2011, recently received USD 6.5 million worth of venture capital. Bina Technologies mainly uses big data to analyze human genetic sequences. Their analysis results serve as fundamental research material for research institutions, clinicians, and other downstream health care service providers.

Bina Technologies' startup team was made up of Ph.D. holders from Stanford University and the University of California, Berkeley. This is a group of avid big data enthusiasts and bioinformatics scientists. They integrated big data with biological sciences. Through Bina's technology in analyzing genetic data, research universities, pharmaceutical companies, and clinicians could use the data to discover rare disease information in genes, which cause cancer, neonatal disease, sickle cell anemia etc.

Søren Nørby, Associate Professor of Human Genetics at the University of Copenhagen explained to reporters that everyone's genes are different, and genetic testing is very important for the diagnosis of patients' diseases. "Such a diagnosis requires scientists to identify the individual's genetic structure and figure out hidden links to the occurrence of diseases in the future, based on the genetic structure. This way, clinical diagnoses can provide more accurate and targeted personalized treatment solutions."

10.4   Leverage on the Charm of Games

Gamification, by definition, is about applying game-design thinking to non-game applications to make them more fun and engaging. Tap into people's natural desire to compete and play, and it results in high levels of engagement.

—Adam Swann

Gamification Comes of Age, in *Forbes* 2012.7.16

"Games in the health care industry are not really about winning or losing," said Michael Fergusson, CEO of game development company Ayogo.

It is human nature to love playing. Compared with boring, dull, or even painful medical care, if attempts can be made to integrate elements of games, software, and hardware users or patients could receive treatment willingly and develop healthy lifestyle habits. This can amplify the effectiveness of wearable health care to a greater extent.

To put it more simply, we have to find ways to integrate the behavioral psychology of play with medical care, so that patients can consciously, actively, and enthusiastically participate in the full health management process. This will help patients improve their health conditions.

10.4.1 Three Major Effects of Gamification

So, exactly what charms would be unleashed when medical health management is gamified?

Firstly, games have the effect of stimulating the inner desires for play and competition in players. If such desires are channeled to certain wearable health care apps, a loyal community can be quickly established. For example, a smart wristband detects your steps or calories burned during daily runs. When you join a dedicated social networking platform and find many friends at the top of the ranks, you will naturally desire to surpass them.

In the future, when human-machine interaction improves further and devices can understand your thoughts, they could loudly alert you whenever an enemy has surpassed you again. This will surely spur you on to work hard and catch up. With such interaction where people try to play catch-up with one another, the goal of running and exercising every day will naturally be achieved. Although this is not an absolutely effective way, it at least plays an auxiliary role.

Secondly, integrating games into some chronic disease management apps and platforms can help patients manage their diseases in their daily lives and adjust their lifestyle habits according to their conditions. For example, Ayogo integrated a game into HealthSeeker, a software designed for diabetics and children who are prone to diabetes. Users can first pick the life goals they wish to accomplish, then complete tasks and accumulate points to earn various badges. As the game was hosted on Facebook, a large social networking platform, there was a large user base that led to the creation of a healthy social circle that was both competitive and interactive.

Ayogo has formed significant partnerships with influential organizations around the world, such as the Joslin Diabetes Center (Harvard Medical School),

Sanofi-Aventis, the University of Southern California Computing Center (Keck School of Medicine), and the Diabetes Hands Foundation.

Social networking games allow people to increase their sense of achievement through competition. Combined with the sharing function provided by the platform, users have more capital to show off. Besides, during the game, users feel good due to the release of more dopamine. This further boosts their desire to continue participating, which releases more dopamine, thus forming a positive feedback loop.

Thirdly, for medical research, the information fed back from gamified medical health management will be more efficient and centralized. This effectively facilitates sample collection and research work. Usually, every game on social networking platforms will have a small community. If medical experts need to study and track the disease management of diabetics, they can enter the community formed for a game software developed specifically for diabetic health management, to collect that information. Such information will certainly be more objective and comprehensive than what is gathered from traditional questionnaires. The information also includes the interactions between patients in their everyday lives and the sharing of their disease management experiences. This information is the most fundamental for researchers.

When activities that people do not usually wish to partake in become interesting, they will naturally appeal to them. To make health management less rigid, one can find ways to gamify it, but the premise is that the game must be easy to play, interesting, able to provide timely feedback and have continuous upgrades.

10.4.2   How to Gamify

Bing Gordon, a partner at KPCB, America's largest venture capital firm, said that CEOs of every startup should be aware of gamification because gamification has become the norm. Certainly, regardless of industry, attempts for gamification can be made. One might even achieve unexpected results.

To get a deeper understanding of how games work in wearable health care specifically, let's take a look at the following companies.

1. Audax Health

In 2013, Audax Health, whose main product was Zensey, a comprehensive medical gamification social and information management platform, received USD 21 million worth of investment.

Zensey is a large and comprehensive internet platform, whose core content is health-themed social networking games designed using behavioral psychology. They use gamification, and other means to get users to live healthier lives. Its functions mainly include sharing on social networks, a challenge, and a motivation system through gamification and one-stop medical information recording and management.

For example, it will award certain points for a healthy lifestyle. Users can share their experiences with patients in the community anonymously, or send their personal medical history information directly to hospitals or family members.

Audax Health believed that the consumers themselves can and should take on active roles in their health management. Through the Zensey platform, Audax Health helps users develop personalized lifestyle plans, communicate with others, track the progress of their goals, and earn rewards for achieving their goals of adopting a healthy lifestyle.

2. Mango Health

Those who do not like taking medication will not take the initiative to do so. But if they do not take medication according to the doctor's advice, their recovery will be impacted. There is thus an app called Mango Health, which specifically reminds people to take their medication.

Mango Health owns a relatively complete drug database. Users only need to add the information of the medication they need to take, and Mango Health will remind them to take the medication and share information related to the drug through in-app reminders and push notifications.

Within Mango Health, the app rewards users with some points based on whether the users have taken their medication. Users' levels will increase with the number of points, and there will be various tangible rewards at different levels (for example, there will be USD 5 worth of Target supermarket rewards when users reach level three, and USD 25 worth of rewards at level five).

One of the more interesting aspects of the Mango Health app is that when a reminder for taking medication has been set, if the user does not enter the app through Mango's notification and check "I have taken the medication" at the scheduled time, Mango will mark a slash on the user's record to indicate that medication was not taken. Furthermore, the app does not have any other openings for users to make up for it. In this way, Mango Health makes users take on an active role, instead of a passive one, in taking their medication.

Mango Health complies with HIPAA standards and aims to improve compliance rates of medication consumption. It provides consumers with an interesting new way to manage medication and nutritional supplements to achieve the goal of health management. Users can receive reminders, so they will not miss any dose, the scheduled interval for taking medication, and the opportunity to earn points to unlock real-world tangible rewards.

3. EveryMove

EveryMove mainly collaborates with insurance companies and health organizations to encourage users to take the initiative to exercise and stay active. For instance, users can choose to provide insurance companies or other health institutions with their healthy lifestyle information collected by various health applications and devices on the EveryMove platform to obtain corresponding rewards. This plan is free for users. Rewards are provided by brands, employers, and health planners.

The founder said that he hoped to help people maintain healthy lifestyles through this positive feedback method. EveryMove collects fees when users redeem rewards, or when health organizations organize activities.

By rewarding people for their healthy living behaviors, EveryMove can improve the health of users on the one hand, and reduce expenditures on medical insurance on the other. For insurance companies, collaborating with EveryMove helps to boost their reputation among members of the public.

In 2012, EveryMove received USD 2.6 million worth of financing. Investors included venture capitalists and big health plan providers. This startup company has attracted many medical providers because it aims to find ways to improve consumer relations and encourage consumer behavior.

4. CogCubed

CogCubed, founded in 2012, was inspired by the Sifteo gaming platform. Its creator Kurt Roots, together with his wife Monika Heller, hoped that CogCubed could help diagnose ADHD (attention deficit hyperactivity disorder) and other disorders among children and adults.

The first game designed by CogCubed, called Groundskeeper, can measure behavioral data when a patient is playing the game, so it can provide accurate data for professionals to analyze whether the patient has ADHD. Its objective is to draw the attention of patients with ADHD through the game, which would thus solve the problems of inability to measure the behaviors of ADHD patients and inaccuracies in data collected in the past.

5. GymPact

GymPact was established in 2013. It was a mobile application that provided supervision and management of fitness plans. Users set up fitness plans on GymPact and placed minimum bets of USD 5, they then used fitness data collected by Fitbit and other wearable devices as the qualifying criterion. Those who failed to achieve their goals would default on their bets, while those who achieved their goals would be rewarded. These reward funds came from the pool of bets from the "Losers." The game also hoped to make everyone honest. It allowed users to vote to upload photos and other evidence of healthy living.

Gamification of health care has become a rising trend, but any company that applies the concept of gamification to health care faces the additional challenge of protecting patient privacy. The participating insurance companies, hospitals, and other medical service providers are not allowed to disclose the personal information of patients at will, and this requires countries to intervene in this area at the legislative level.

Index

Dr. Kevin Chen, or Chen Gen, is a renowned science and technology writer and scholar, postdoctoral scholar at Boston University, and an invited course professor at Peking University. He has served as special commentator and columnist for *The People's Daily*, CCTV, China Business Network, SINA, NetEase, and many other media outlets. He has published several monographs involving numerous domains, including finance, science and technology, real estate, medical treatment, and industrial design. He has currently taken up residence in Hong Kong.

CPSIA information can be obtained
at www.ICGtesting.com
Printed in the USA
LVHW092154070521
686849LV00001B/2